Macromolecular Symposia

Symposium Editors: B. Voit, F. Böhme, H.-J. Adler

Editor: I. Meisel
Associate Editors: K. Grieve, C.S. Kniep, S. Spiegel

Executive Advisory Board: M. Antonietti, M. Ballauff, H. Höcker,
S. Kobayashi, K. Kremer, T.P. Lodge,
H.E.H. Meijer, R. Mülhaupt,
A.D. Schlüter, H.W. Spiess, G. Wegner

177

pp. 1–191

January 2002

Macromolecular Symposia publishes lectures given at international symposia and is issued irregularly, with normally 14 volumes published per year. For each symposium volume, an Editor is appointed. The articles are peer-reviewed. The journal is produced by photo-offset lithography directly from the authors' typescripts.

Further information for authors can be obtained from:

Editorial office "Macromolecular Symposia"

Wiley-VCH

P. O. Box 10 11 61, 69451 Weinheim,

Germany

or for parcel and courier services: Pappelallee 3, 69469 Weinheim,

Germany

Tel. +49 (0) 62 01/6 06-2 38 or -5 81; Fax +49 (0) 62 01/6 06-3 09 or 5 10; macromol@wiley-vch.de

http://www.wiley-vch.de/home/macrosymp

Suggestions or proposals for conferences or symposia to be covered in this series should also be sent to the Editorial office at the address above.

Macromolecular Symposia:
Annual subscription rates 2002 (print only or online only)*

Germany, Austria € 1198; Switzerland SFr 1968; other Europe € 1198; outside Europe US $ 1448.

Macromolecular Package, including Macromolecular Chemistry & Physics (18 issues), Macromolecular Rapid Communications (18 issues), Macromolecular Theory & Simulations (9 issues) is also available. Details on request.

* For a 5 % premium in addition ot **Print Only** or **Online Only** Institutions can also choose both print and online access.

Packages including Macromolecular Symposia and Macromolecular Materials & Engineering are also available. Details on request.

Single issues and back copies are available. Please, inquire for prices.

Orders may be placed through your bookseller or directly at the publishers: WILEY-VCH Verlag GmbH, P. O. Box 10 11 61, 69451 Weinheim, Germany, Tel.: (0 62 01) 6 06–0, Telefax (0 62 01) 60 61 17, Telex 46 55 16 vchwh d. E-mail: subservice@wiley-vch.de

Macromolecular Symposia (ISSN 1022-1360) is published with 14 volumes per year by WILEY-VCH Verlag GmbH, P.O. Box 10 11 61, 69451 Weinheim, Germany. Air freight and mailing in the USA by Publications Expediting Inc., 200 Meacham Ave., Elmont, NY 11003. Periodicals postage pending at Jamaica, NY 11431. POSTMASTER: send address changes to Macromolecular Symposia, Publications Expediting Inc., 200 Meacham Ave., Elmont, NY 11003.

8th Dresden Polymer Discussion
Synthesis of Defined Polymer Architectures

held at
Meißen near Dresden, Germany
April 23 to 26, 2001

Organized by

Institut für Polymerforschung Dresden e.V.
Technische Universität Dresden,
Institut für Makromolekulare Chemie und Textilchemie

Symposium Editors and Scientific Committee

B. Voit
F. Böhme
Institut für Polymerforschung Dresden e.V.
Hohe Str. 6
01069 Dresden
Germany

and

H.-J. Adler
Technische Universität Dresden,
Institut für Makromolekulare Chemie und Textilchemie
01062 Dresden
Germany

Organization:

K. Wustrack

The 8th Dresden Polymer Discussion was sponsored by DFG, Sächsisches Staatsministerium für Wissenschaft und Kunst, ACORDIS Industrial Fibres GmbH, BAYER AG, BASF AG, and Creavis GmbH.
The meeting was linked to the activities of DFG-Sonderforschungsbereich 287 "Reactive polymers in inhomogeneous systems, in melts and at surfaces"

Contents of Macromol. Symp. 177

8th Dresden Polymer Discussion: Synthesis of Defined Polymer Architectures
Meißen near Dresden (Germany), 2001

Preface
B. Voit

* The asterisk indicates the name of the author to whom inquiries should be
addressed

Author Index

Preface

Dresden Polymer Discussions have been organized by the Institute of Macromolecular Chemistry and Textile Chemistry of the Dresden University of Technology and the Institute of Polymer Research Dresden on a biennial basis since 1986. While the meetings have been dedicated to different specific fields of polymer science and the conference venue had to be relocated twice, a constant and crucial feature of the series of meetings has always been intensive and inspiring discussion within a limited circle of (mostly invited) participants from leading research groups and industry. This meeting has developed over the years to a well established high quality Gordon-type conference and the specific intimate character of the meeting was supported by the choice of the conference venue, the Evangelische Akademie, in the thoroughly restored former St. Afra Monastery in Meissen, an old town about 25 km from Dresden.

The 8th Dresden Polymer Discussion (April 23 to 26, 2001) dealt with one of the most topical fields of research in polymer synthesis, the synthesis of defined polymer structures. New and interesting results were presented concerning controlled synthetic strategies (ionic, cationic, radical, metallocene catalyzed), specific polymer architectures (block, graft, stars, dendritic), and supramolecular structures. One section of the conference was dominated by 'living' polymerization techniques with special focus on controlled radical polymerization. It was one of the

Participants of the 8th Dresden Polymer Discussion

rare cases to have Kris Matyjaszewski, Mitsuo Sawamoto, Virgil Percec and Craig Hawker together in one meeting with enough time for discussion. In total, the scientific program included 28 lectures and 20 posters, which were presented and intensively discussed by almost 40 permanent (invited) participants from eight countries, as well as about the same number of research fellows, students and postdocs from the organizing institutes.

Finally, it should not be forgotten that the success of the meeting is to a considerable extent also due to the sponsorship of the German Science Foundation (DFG), Saxon Ministry for Science and the Fine Arts (SMWK), Acordis Industrial Fibres GmbH, BASF AG, Bayer AG, and Creavis Gesellschaft für Technologie und Innovation mbH.

A selection of the oral papers will presented now in this *Macromolecular Symposia* volume.

Dresden, September 2001   *B. Voit*
on behalf of the Symposium Editors

Graft Copolymers by Atom Transfer Polymerization

Hans G. Börner and Krzysztof Matyjaszewski[*]

Center for Macromolecular Engineering, Department of Chemistry, Carnegie Mellon University, 4400 Fifth Avenue, Pittsburgh, PA 15213, USA

Summary: Various graft copolymers have been prepared by atom transfer radical polymerization (ATRP) using both "grafting through" and "grafting from" approaches. The synthesis and some properties are reviewed.

Introduction

Graft copolymers represent valuable polymeric materials, since a variety of molecular parameters can be varied (as shown in Scheme 1): i) main and side chain polymer type, ii) degree of polymerization and polydispersities of main and side chain, iii) graft density (average spacing in-between the side chains) and iv) distribution of the grafts (graft uniformity). Using special polymerization techniques tailor-made graft copolymers can be afforded according to specific needs.

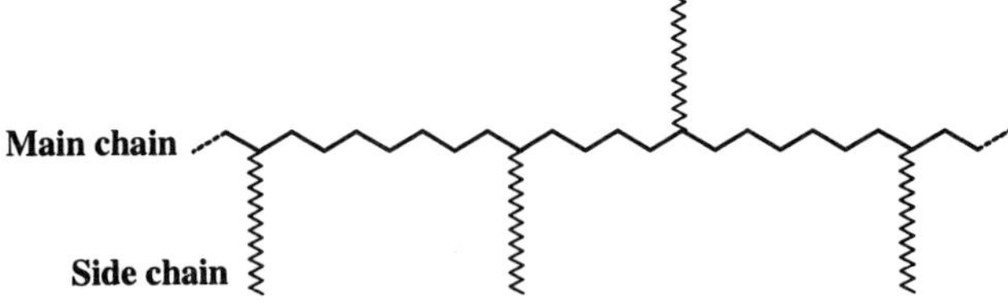

Scheme 1: Graft copolymer.

By controlling the molecular parameters, one can obtain impact resistant materials by combining a hard polymer backbone with soft polymer side chains; thermoplastic elastomers, where a soft polymer backbone is grafted with hard polymer segments; or amphiphilic copolymers for applications as hydrogels, stabilizers, surface-modifying agents, dispersants, emulsifiers and compatibilizers in polymer blends, etc.[1,2].

Three main methods can be used to prepare graft copolymers: i) "grafting through", where a macromonomer is copolymerized with a low molecular weight comonomer; ii) "grafting from"; in this case a macroinitiator with predetermined initiation sides is used to initiate the polymerization of a second monomer; iii) "grafting onto", where an end-functional or a living polymer with reactive end-group is coupled with functional groups located on another polymer.

Until recently, well-defined graft copolymers could be obtained only by using ionic polymerizations or the macromonomer route. Due to the lack of control of molecular weight and polydispersity, the free-radical polymerization methods (FRP) lead to poorly defined polymers, very often accompanied by a certain amount of homopolymer and crosslinked material[1,3]. Free-radical techniques though have become available for the preparation of well-defined polymers. The most important types of controlled/"living" radical polymerizations are: i) stable free radical polymerization (SFRP), which usually employs nitroxyl radicals[4,5]; ii) atom transfer radical polymerization (ATRP), where complexes of transition metals such as copper[6,7], ruthenium[8], iron[9,10] or nickel[11,12], are utilized in conjunction with alkyl halides; iii) reversible addition-fragmentation chain transfer polymerization (RAFT), where dithioesters coupled with a free radical initiator induce polymerization[13]. So far only SFRP[14] and ATRP[15] have been used to prepare graft copolymers by both "grafting through" and "grafting from" methods.

ATRP has been applied in the synthesis of several well-defined graft copolymers. Two main approaches have been adapted: i) the "grafting through" technique, i.e. a well-defined macromonomer was synthesized in the first step by ATRP, followed by its free-radical or controlled radical copolymerization with a low molecular weight comonomer (Scheme 2); ii) the "grafting from" technique, i.e. a polymer possessing ATRP-active halide atoms as side groups was used as a macroinitiator in the ATRP of a second monomer. Generally, these macroinitiators can be obtained directly by copolymerizing a monomer containing ATRP-initiator group using a polymerization mechanism that is different from ATRP (FRP, SFRP, ring opening polymerization (ROMP)) or by copolymerization of a monomer carrying a precursor group that can be transformed to an ATRP-initiating group (Scheme 3).

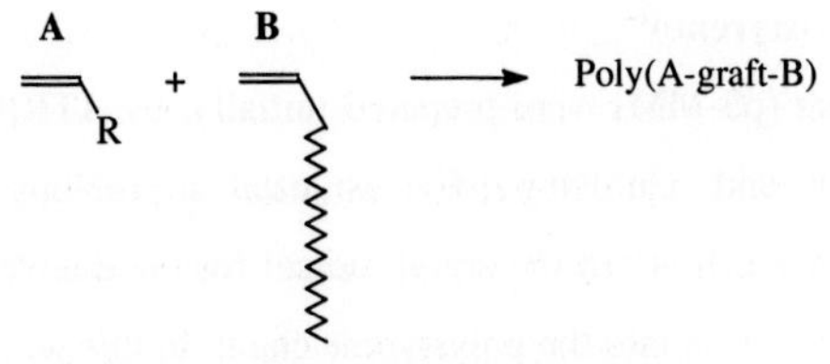

Poly(A-graft-B)

Scheme 2: "Grafting through" approach.

Functional Group Transformation

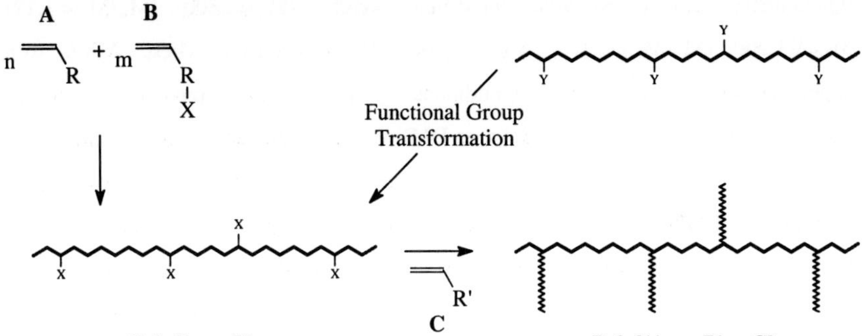

Poly(A-*co*-B)

Poly[(A-*co*-B)-*g*-C]

Scheme 3: "Grafting from" approach.

The present paper gives an overview over the graft copolymer materials prepared by copper-mediated ATRP.

1 Graft copolymers by the "grafting through" approach

The "grafting through" approach is the most common route to controlled graft copolymers[16,17]. The copolymerization of macromonomers (MM) with low molecular weight comonomer allows good control of the polymer side chain parameters (DP, M_w/M_n) by using living or controlled polymerization methods for the MM synthesis. The polymer main chain parameters (DP, M_w/M_n) can be controlled by a living copolymerization of the MM with a low molecular weight comonomer. The average side chain density is determined by the reactivity ratio of the MM in the copolymerization and by the ratio of the MM in the feed. The spacing distribution of the side chains is the most crucial parameter to control[18]. It is influenced by the diffusion differences of MM in comparison to the low molecular weight comonomer[19], by the inherent reactivity of MM and comonomer[20] and by the potential incompatibility of the side chain polymer with the main chain polymer[21].

1.1 Poly(N-vinyl pyrrolidinone-*graft*-styrene)[22]

Well defined polystyrene macromonomers (pS-MM) were prepared initially, by ATRP using vinyl chloroacetate as initiator and Cu[dNbpy]$_2$Cl as catalyst (dNbpy: 4,4'-Di(5-nonyl)-2,2'-bipyridine). The low reactivity of the styryl radical for the double bond of vinyl acetate suppresses its incorporation into the polystyrene chain. In this way vinyl acetate end-functionalized pS-MM was obtained having low polydispersity, high functionality and controllable molecular weight ($M_n = 5800$ ($M_w/M_n = 1.12$); $M_n = 11,900$ ($M_w/M_n = 1.15$); $M_n = 15,900$ ($M_w/M_n = 1.18$)). These MMs were copolymerized with N-vinylpyrrolidinone (NVP) using conventional radical polymerization with AIBN as initiator in DMF as solvent for both main and side chain polymer. The amount of incorporated MM (average graft density) depends on the MM ratio in the feed, but also on the MM length (decrease in graft density with increase in M_n (MM) by constant feed ratio) (Table 1). The synthesized material can be described as a thermoplastic hydrogel. It follows the concept of thermoplastic elastomers, since the hydrophilic pNVP main chains are swellable in polar solvents (e.g. water) and lead to hydrogel formations. Dissolution in water is prevented by the physically crosslinked, hydrophobic pS side chains. Microphase separation was confirmed by differential scanning calorimetry (DSC) analysis since two glass transition temperatures can be observed (e.g. $T_{g,1} = 74.79$ (pNVP); $T_{g,2} = 111.3$ (pS); Sample Table 1, entry 4).

Table 1: Synthesis of poly(N-vinyl pyrrolidinone-*graft*-styrene).

pS-MM M_n (M_w/M_n)	MM incorporated w% (feed)	Copolymer M_n (M_w/M_n)	Grafts / Chain	Equal. Content of water (%)
	34 (40)	316000 (5.90)	18.6	85
5800 (1.12)	13 (20)	219000 (2.50)	4.9	92
	8 (10)	185000 (1.80)	2.5	97
11900 (1.15)	40 (50)	65700 (1.58)	2.2	82

1.2 Poly(methyl methacrylate-*graft*-dimethylsiloxane)[21]

The polydimethylsiloxane (PDMS) macromonomer was synthesized by anionic ring-opening polymerization and subsequently end-functionalized with 3-methacryloxypropyl-dimethylchlorosilane (PDMS-MA: $M_n = 2370$; $M_w/M_n = 1.25$; $F = 1.0$)[23]. The macromonomer was copolymerized with methyl methacrylate employing various polymerization mechanisms like FRP, RAFT and ATRP under diverse conditions (Table 2). The inverse reactivity ratio which is the reciprocal value of $r_{(comonomer)}$ as determined by Jaacks method[24] is equivalent to the relative reactivity of the MM ($1/r_{(comomomer)} = k(comonomer-MM)/ k(comonomer-comonomer)$). $1/r_{MMA}$ was estimated in order to gain some insight about the side chain distribution in the synthesized graft copolymer.

The copolymer synthesized by conventional free radical polymerization (FRP) shows a first order heterogeneity; i.e. the chemical composition of different polymer molecules is different. This can be attributed to the short lifetime of the active growing polymer chain (~ s) and the shift of comonomer feed by faster incorporation of the low molecular weight comonomer, as indicated by the reactivity ratio of $1/r_{MMA} = 0.34$ (Table 2, entry 1). Using RAFT or ATRP, all chains grew simultaneously with the same chemical composition. According to the reactivity ratio ($1/r_{MMA} \neq 1$) the feed changes during polymerization. This results in polymers having chains of constant chemical composition, but a heterogeneous composition along each polymer chain is evident; i.e. the so-called second order heterogeneity is apparent. The obtained graft copolymer showed a gradient in side chain density (Table 2, entry 2-5). Using ATRP with a PDMS-BiB macroinitiator in semi-bulk conditions the reactivity ratio approached the ideal inverse reactivity ratio of 1. This indicates incorporation of both macromonomer and comonomer with comparable ratios and control over side chain distribution.

Table 2: Synthesis of poly(methyl methacrylate-*graft*-dimethylsiloxane).

		Conditions	Xylene (w%)	$1/r_{MMA}$ [a]	Distribution
1	FRP	[1] AIBN	31	0.34	
2	RAFT	[1,2] BPO	31	0.67	
3	RAFT	[1,2] AIBN	31	0.57	
4	ATRP	[1,3] EBiB, 75°C	31	0.73	
5	ATRP	[1,3] EBiB, 90°C	31	0.81	
6	ATRP	[1,3] PDMS-BiB	3	0.85	

[1] MMA : PDMS-MA = 95 mol% : 5 mol%
[2] M : Cumyl dithiobenzoate : I = 300 : 1 : 0.5
[3] M : Cu[dNbpy]$_2$Cl : In = 300 : 1 : 1
[a] determined by Jaacks method[24].

1.3   Poly(methyl methacrylate-*graft*-lactide)[20]

Two different well-defined polylactide (pLA) macromonomers were prepared using hydroxyethyl methacrylate (MA) or hydroxyethyl acrylate (A) as initiator in the anionic ring-opening polymerization of lactide (pLA-MA, pLA-A; Table 3).

Table 3: PLA-Macromonomers.

	M_n [+]	M_w/M_n [+]	Functionality
pLA-MA	2800	1.16	1.0
pLA-A	2700	1.22	0.99

[+] determined by GPC (THF; pMMA-standards)

The pLA-macromonomers were copolymerized with methyl methacrylate using ATRP. Since the slow ATRP reaction makes the diffusion control effect negligible and the polylactide side chains are compatible with the pMMA main chain, the reactivity of the MM is mainly determined by the inherent reactivity of the head group. The strong dependence of the reactivity ratio on the head group of the MM is shown in Table 4. The copolymerization of the oxyethylene methacrylate terminated MM (pLA-MA) with MMA results in a reactivity ratio of $1/r_{MMA} = 1.85$. On the other hand, $1/r_{MMA} = 0.61$ can be determined in the copolymerization of oxyethylene acrylate terminated MM (pLA-A) with MMA. Control over the side chain distribution can be achieved by polymerizing a mixture of both pLA-MA and pLA-A in a 50 : 50 w% ratio. The determined inverse reactivity ratio $1/r_{MMA} = 0.93$ is close to the ideal value of 1 and indicates a MM and comonomer incorporation in the polymer with comparable ratios. Furthermore, the obtained graft copolymer showed a molecular weight of $M_n = 55,000$, a low polydispersity index ($M_w/M_n = 1.16$), and in average 3.5 grafts per 100 MMA repeat units.

Table 4: Copolymerization of pLA-MM with MMA.

Comonomers [a]	$1/r_{MMA}$ [b]
MMA + pLA-MA[*]	1.85
MMA + pLA-A[**]	0.61
MMA + 0.5xpLA-MA[*] + 0.5xpLA-A[**]	0.93

[a] Conditions: xylene : diphenylether = 1 : 1; 90 °C; Cu[dNBpy]$_2$Cl; 3.5 mol% MM
M : In : Cat = 300 : 1 : 1; 90% monomer conversion (GC).
[b] determined by Jaacks method[24].
[*] oxyethyl methacrylate head group.
[**] oxyethyl acrylate head group.

1.4 Poly(*n*-butyl acrylate-*graft*-methyl methacrylate)[25,26]

The copolymerization of methacryloyl end-functionalized poly(methyl methacrylate) macromonomer (pMMA-MM) obtained by GTP[3,19] with *n*-butyl acrylate was elucidated by comparing FRP and ATRP as two different polymerization mechanisms.

Model studies were followed, to determine differences in reactivity ratios of low molecular weight MMA and *n*-BuA. The determined reactivity ratios in the copolymerization of MMA with *n*-BuA were similar in both conventional and controlled radical polymerizations. This indicates that the selectivity of the corresponding radicals is independent of the polymerization mechanism.

In copolymerization of pMMA-MM with *n*-BuA the molecular weight difference between both monomers caused a lower mobility of the MM, which appeared strongly at higher MM concentrations.

In FRP the relative reactivity of the MM decreased with increase in MM concentration, due to the diffusion control. In ATRP the diffusion control effect can be suppressed by extending the time scale of the polymerization. This leads at comparable monomer concentrations to much higher relative reactivity of the MM which is closer to the value of MMA since it is the low molecular weight model for the MM end-group. Due to the controlled nature of the ATRP reaction, the apparent molecular weights of the graft copolymer increased linearly with conversion and the polydispersities were lower than those obtained by FRP. However the polydispersity indices increased at high conversions, indicating side reactions such as termination or chain transfer.

The products of the pMMA-MM copolymerization with *n*-BuA using three different polymerization mechanisms such as FRP, ATRP and living anionic polymerization were analyzed by 2D chromatography[27] (HPLC (p*n*BuA elution under critical conditions (LACCC)[28])-GPC). The results are presented in Table 5. It has been found that the polymerization mixture composition depends strongly on the mechanism. Four main compounds could be identified: i) the poly(*n*BuA-*graft*-MMA) graft copolymer, which is the desired product; ii) poly(*n*BuA-*star*-MMA) that has only one pMMA graft incorporated in the p*n*BuA chain; iii) the remaining pMMA-MM and iv) p*n*BuA-homopolymer. In contrast to FRP that led to 63% graft copolymer, 17% star copolymer and 9% p*n*BuA homopolyer, the ATRP gave > 90% of the desired graft copolymer and less than 1% p*n*BuA homopolymer (Table 5). Using living anionic polymerization techniques the lowest polydispersity value of the graft copolymer was obtained ($M_w/M_n = 1.3 < 1.7$ (FRP) < 1.8 (ATRP)). However there was 31% MM left and the distribution of the grafts in the graft copolymer is probably not random but strongly tapered, due to the reactivity differences of the comonomers that occurs strongly in anionic polymerization.

Table 5: Composition of polymerization mixture determined by 2D chromatography.

Mechanism	Percentage of compound after polymerization ($M_{n,app}$; M_w/M_n)			
	Graft copolymer p(nBuA-g-MMA)	Star copolymer p(nBuA-$star$-MMA)	Macromonomer pMMA-MM	Homopolymer pnBuA
FRP [a]	63 (54800; 1.7)	17	8[*]	9
ATRP [b]	91 (72600; 1.8)	6	1 (5900; 2.1)[*]	1
anionic [c]	40 (86000; 1.3)	18	31 (7800; 1.8)[**]	5

[a] AIBN; 60 °C; butyl acetate as solvent [3].

[b] methyl • -bromopropionate, Cu[dNbpy]$_2$Br, Cu^0; 90 °C; diphenyl ether as solvent[26].

[c] ethyl • -lithioisobutyrate, 2:1 complex of triisobuthylaluminium and CsF; -78 °C; toluene as solvent[29,30].

[*] methacryloyl head group.

[**] acryloyl head group.

1.5   Poly(n-butyl acrylate-$graft$-ethylene)[31]

Methacryloyloxy end-functionalized polyethylene macromonomer (pE-MM) with molecular weight of $M_n \sim 10000$ and polydispersity of $M_w/M_n < 1.04$ were prepared using a Pd catalyst[32-34]. Since the polyethylene was highly branched (106-92 branches per 1000 C) and amorphous the macromonomer was soluble in common solvents such as hexanes, THF, chloroform and also in n-butyl acrylate. Using ATRP techniques the MM could be copolymerized with n-butyl acrylate. The inverse reactivity ratio, determined by Jaacks method, was found to be highly dependent on the initial concentration of MM (Table 6). At low concentration and consequently decreased viscosity of the reaction mixture, the MM was consumed faster than the comonomer, owing to the higher reactivity of the methacryloyloxy group. This was manifested by $1/r_{nBuA} = 2.38$. At higher pE-MM concentrations (Table 6, entry 2-3), the viscosity of the solution increased lowering the mobility of the macromonomer and causing a decrease in the apparent reactivity. Nevertheless the polymerization was in all cases controlled which was demonstrated by the low polydispersity values (Table 6).

Table 6: Poly(n-butyl acrylate-g-ethylene) by ATRP of polyethylene macromonomer with n-butyl acrylate.

	$(Comp.)_0$ [a] (wt%)	M_n ($\times 10^{-3}$)	M_w/M_n	Comp. [b] (wt%)	$1/r_{nBuA}$ [c]
1	25.0	115	1.6	34.7	2.38
2	50.0	86	1.4	59.2	1.72
3	71.4	70	1.4	70.2	0.94

[a] Initial pE-MM weight % in monomers.

[b] Polyethylene weight % in the graft copolymer determined by the conversion of n-BuA and pE-MM.

[c] Reactivity ratios determined by Jaacks method.

Polymerization conditions:

1) n-BuA : pE-MM : DMDBHD (dimethyl 2,6-dibromoheptanedioate) : CuBr : PMDETA (N,N,N',N'',N''-pentamethyldiethylenetriamine) = 1669:6.0:1:20:20, $[nBuA]_0 = 4.92$ mol/L; t = 15 h.

2) n-BuA : pE-MM : DMDBHD : CuBr : PMDETA = 546:6.3:1:10:10, $[nBuA]_0 = 3.48$ mol/L; t = 6.7 h.

3) n-BuA : pE-MM : DMDBHD : CuBr : PMDETA = 234:7.0:1:10:10, $[nBuA]_0 = 2.11$ mol/L; t = 15 h.

2 Graft copolymers by the "grafting from" approach

The "grafting from" approach uses a macroinitiator (MI) that carries initiation groups to start a polymerization reaction of a monomer. The synthesis of the MI allows the control of the main chain parameters (DP, M_w/M_n) as well as the pre-determination of side chain density and side chain distribution. The main difficulty in this approach is to ensure a uniform propagation of the side chains and furthermore to suppress chain coupling reactions.

The method is not restricted to linear MI, particles[35,36] and surfaces[37,38] were also successfully functionalized with initiating groups and used as MI to get access to e.g. inorganic organic hybrid materials.

2.1   Macroinitiators leading to graft copolymers with low graft density

Table 7: Graft copolymers from polyolefin macroinitiators.

	Macroinitiator			Monomer	Graft copolymer			
	Polymer	M_n $(\times10^{-3})$	M_w/M_n		Monomer (wt.-%)	M_n $(\times10^{-3})$	M_w/M_n	T_g (°C)
1	PIB[1]	200	2.1	Sty[2]	65	435	2.4	-64/93
		"	"	Sty	18	180	2.5	-57/ -
		"	"	MMA[3]	68	300	3.5	-70/100
		"	"	MMA	15	200	3.5	-59/110
		108	2.3	IBuA[4]	21	181	2.5	-10
2	PEGM[5]	-	-	Sty	69	-	-	-15/108
				MMA	80	-	-	- /125
3	CSPE[6]	14.9	2.3	Sty	-	85.6	1.8	-10/87
				MMA	-	26.3	1.8	-2
4	PVC[7]	47.4	2.7	Sty	80	99.5	3.7	80
				MA	50	57.7	2.4	21
				MMA	60	83.6	4.9	111
				n-BuA[8]	65	81.4	2.4	-19

[1]PIB = poly(isobutylene-co-p-methylstyrene-co-p-bromomethylstyrene), [2]Sty = styrene, [3]MMA = methyl methacrylate, [4]IBuA = isobornyl acrylate, [5]PEGM = poly(ethylene-co-glycidyl methacrylate), [6]CSPE = chlorosulfonated polyethylene, [7]PVC = poly(vinyl chloride-co-vinyl chloroacetate), [8]BuA = n-butyl acrylate.

2.1.1   Graft copolymers of poly(isobutene) [39-41]

Thermoplastic elastomers were prepared starting from commercially available partially brominated poly(isobutene-co-p-methystyrene) with 1.2 mole-% benzylic bromine (Exxpro[TM] elastomer), which was used as a macroinitiator to initiate ATRP of styrene[39-41], methyl methacrylate[41] and isobornyl acrylate[39] (Table 7, entry 1). The molecular weights of the graft copolymers increased linearly with conversion. Above a certain ratio of comonomer in the graft copolymer the glass transition temperatures of both backbone and grafts could be observed in DSC analysis, indicating microphase separation (Table 7). Combining soft (low T_g) polymers with hard glassy polymers lead to thermoplastic elastomers or impact strength modified plastics, if microphase separation takes place.

2.1.2 Graft copolymers from polyethylene[42]

Free radical polymerization methods were mainly used for the synthesis of graft copolymers from non-polar olefin backbones. "Grafting from" via irradiation initiation was mostly employed[43]. More recently, metallocene[44] and nitroxide-mediated[33] methods have been used to grow styrene grafts from ethylene and propylene backbones respectively.

Commercially available poly(ethylene-co-glycidyl methacrylate) was converted into a suitable ATRP macroinitiator by reaction of the epoxide groups with chloroacetic acid or 2-bromoisobutyric acid. From this pendant functionalized macroinitiator styrene and methyl methacrylate was grafted using ATRP conditions (Table 7, entry 2). The amount of incorporated comonomer could be controlled by the monomer conversion. The linear increase of the molecular weight with conversion and low polydispersities of $M_w/M_n < 1.4$ of the cleaved side chains indicated a controlled growing reaction. DSC analysis of purified samples showed two glass transitions, which were characteristic for polyethylene and polystyrene or poly(methyl methacrylate) segments respectively (Table 7).

2.1.3 Graft copolymers of chlorosulfonated polyethylene

Chlorosulfonated polyethylene (CSPE) is an elastomer that contains chlorine atoms and sulfonyl chloride groups randomly distributed on a polyethylene chain. Sulfonyl chloride groups are known to be good initiators for ATRP[45]. Therefore, CSPE can be used as macroinitiator in ATRP to produce graft copolymers[46,47]. Table 7 (entry 3) shows the results of CSPE grafting with styrene and MMA by ATRP.

2.1.4 Graft copolymers of poly(vinyl chloride)[48]

Graft copolymers of poly(vinyl chloride) with styrene and (meth)acrylates have been prepared by ATRP using a statistical copolymer of vinyl chloride with 1 mole-% vinyl chloroacetate as macroinitiator (Table 7, entry 4). For *n*-butyl acrylate polymerization the molecular weight of the graft copolymer increased with increasing incorporation of the monomer, while the glass transition temperature decreased. All graft copolymers prepared displayed only one glass transition temperature, indicating no sufficient microphase separation (Table 7).

2.1.5 Graft copolymers of polysiloxanes[49]

Graft copolymers composed in part of polysiloxanes can be prepared by using a hydrosilation reaction to introduce 2-(4-chloromethylphenyl)ethyldimethylsilane to the vinyl groups of poly(dimethyl-co-vinylmethylsiloxane). The obtained macroinitiator with pendant benzyl chloride functionalities can initiate the polymerization reaction of styrene under ATRP conditions. The resulting graft copolymer is an inorganic/organic composite material, consisted out of an inorganic backbone and organic grafts.

2.2 Macroinitiators leading to graft copolymers with high graft density[50-52]

The "grafting from" method is also a convenient route to polymer systems with high graft density. Well-defined, macroinitiators can be prepared that have ATRP initiation groups as side groups in each repeat unit. These can be accessed by transformation of a well-defined MI precursor (poly(2-trimethylsilyloxyethyl methacrylate; pHEMA-TMS) which has already been prepared by ATRP with 2-bromopropionyl bromide into poly(2-(2-bromopropionyloxy)ethyl methacrylate) (pBPEM)[50,52].

Molecular brushes can be prepared with homopolymer side chains of polystyrene and various polyacrylates with good control as indicated by the low over all polydispersity of the molecular brushes and the low polydispersities of the cleaved side chains $M_w/M_n = 1.2$ [52,53]. The resulting graft copolymers provide sufficient steric repulsion between the side chains to force the backbone into an extended worm-like conformation. The molecules are large enough to be resolved individually on mica surface using atomic force microscopy (AFM) in tapping mode.

A variety of different molecular parameters can be controlled using this method (Scheme 4): i) aspect ratio (l - axial length versus r - molecule radius); ii) side chain bulkiness (b - methyl < n-buthyl < t-buthyl acrylate side chains); iii) main chain architecture (block copolymer); iv) graft density (• - spacing in between the grafts); v) graft density gradient (• •); vi) side chain architecture (block copolymer side chains). Thus, ATRP has provided the means to prepare densely grafted macromolecules with more well-defined dimensions without the subsequent fractionation at size exclusion columns necessary.

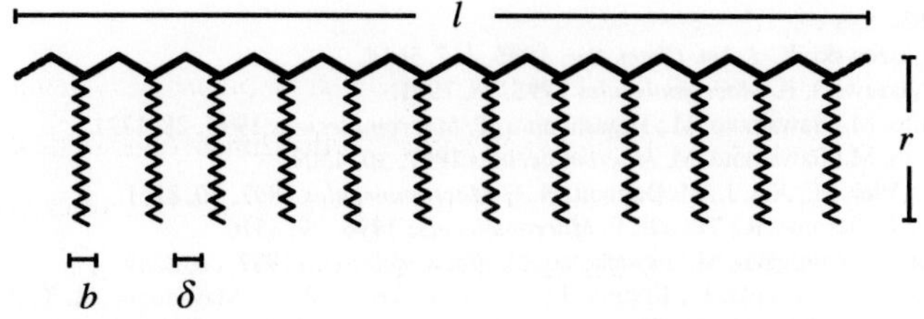

Scheme 4: Molecular parameters in a molecular brush.

Densely grafted polymer systems are also accessible by "grafting through" techniques in combination with ATRP[54]. However, the homopolymerization of macromonomers presents a limitation as far as high targeted DP is concerned.

Conclusions

Atom transfer radical polymerization (ATRP) was used to synthesize a variety of graft copolymers, differing in main and side chain polymer types. Using "grafting through" and "grafting from" methods, the molecular parameters such as the degree of polymerization of and the polydispersity of main and side chain polymer, graft density and graft distribution could be controlled.

Acknowledgment

Beers, K. L.; Gaynor, S. G.; Hong, S. C.; Kickelbick, G.; Kowalewski, T.; Miller, P. J.; Nakagawa, Y.; Pyun, J.; Peterson, M. L.; Paik, H.-j.; Patten, T. E.; Roos, S. G; Shinoda, H.; Teodorescu, M.; Wang, J. S.

This work was financially supported by the National Science Foundation, ATRP/CRP Consortia at CMU and Deutsche Forschungsgemeinschaft (DFG, Germany).

(1)   Rempp, P. F.; Lutz, P. J. In *Comprehensive Polymer Science*; Allen, G., Bevington, J. C., Eds.; Pergamon Press: Oxford, **1989**; Vol. 6; p. 403.
(2)   Pitsikalis, M.; Pispas, S.; Mays, J. W.; Hadjichristidis, N. *Adv. Polym. Sci.* **1998**, *135*, 1.
(3)   Roos, S. G.; Müller, A. H. E.; Kaufmann, M.; Siol, W.; Auschra, C. *ACS Symp. Ser.* **1998**, *696*, 209.
(4)   Georges, M. K.; Veregin, R. P. N.; Kazmaier, P. M.; Hamer, G. K. *Macromolecules* **1993**, *26*, 2987.
(5)   Matyjaszewski, K. *Controlled Radical Polymerizations*; American Chemical Society: Washington,

DC, **1998**; Vol. 685.

(6) Wang, J. S.; Matyjaszewski, K. *J. Am. Chem. Soc.* **1995**, *117*, 5614.

(7) Wang, J. S.; Matyjaszewski, K. *Macromolecules* **1995**, *28*, 7901.

(8) Kato, M.; Kamigaito, M.; Sawamoto, M.; Higashimura, T. *Macromolecules* **1995**, *28*, 1721.

(9) Ando, T.; Kamigaito, M.; Sawamoto, M. *Macromolecules* **1997**, *30*, 4507.

(10) Matyjaszewski, K.; Wei, M.; Xia, J.; McDermott, N. E. *Macromolecules* **1997**, *30*, 8161.

(11) Granel, C.; Dubois, P.; Jerome, R.; Teyssie, P. *Macromolecules* **1996**, *29*, 8576.

(12) Uegaki, H.; Kotani, Y.; Kamigaito, M.; Sawamotao, M. *Macromolecules* **1997**, *30*, 2249.

(13) Chiefari, J.; Chong, Y. K.; Ercole, F.; Krstina, J.; Jeffery, J.; Le, T. P. T.; Mayadunne, R. T. A.; Meijs, G. F.; Moad, K. L.; Rizzardo, E.; Thang, S. H. *Macromolecules* **1998**, *31*, 5559.

(14) Hawker, C. J.; Malmstrom, E. E.; Frechet, J. M. J.; Leduc, M. R.; Grubbs, R. B.; Barclay, G. G. *Controlled Radical Polymerization*; American Chemical Society: Washington, D.C., **1998**; Vol. 685.

(15) Patten, T. E.; Matyjaszewski, K. *Adv. Mat.* **1998**, *10*, 901.

(16) Schulz, G. O.; Milkovich, R. *J. Appl. Polym. Sci.* **1982**, *27*, 4773.

(17) Mishra, M. K. *Macromolecular Design: Concept and Practice - Macromonomers, etc.*; Polymer Frontier International: New York, **1994**.

(18) Meijs, G. F.; Rizzardo, E. *J. Macromol. Sci., Rev. Macromol. Chem.* **1990**, *C30*, 305.

(19) Radke, W.; Müller, A. H. E. *Macromol. Chem., Macromol. Symp.* **1992**, *54*, 583.

(20) Shinoda, H.; Matyjaszewski, K. *Macromolecules* **2001**, *submitted*.

(21) Shinoda, H.; Miller, P. J.; Matyjaszewski, K. *Macromolecules* **2001**, *34*, 3186.

(22) Matyjaszewski, K.; Beers, K. L.; Kern, A.; Gaynor, S. G. *J. Polym. Sci., Part A: Polym. Chem.* **1998**, *36*, 823.

(23) Schunk, T. C.; Long, T. E. *J. Chromatogr.* **1995**, *A692*, 221.

(24) Jaacks, V. *Macromol. Chem.* **1972**, *161*, 161.

(25) Roos, S. G.; Müller, A. H. E.; Matyjaszewski, K. *ACS Symp. Ser.* **2000**, *768*, 361.

(26) Roos, S. G.; Müller, A. H. E.; Matyjaszewski, K. *Macromolecules* **1999**, *32*, 8331.

(27) Müller, A. H. E.; Roos, S. G.; Schmitt, B. *Polym. Prepr. (Am. Chem. Soc., Div. Polym. Chem.)* **1999**, *40*, 140.

(28) Pasch, H.; Much, H.; Schulz, G.; Gorshkov, A. V. *LC GC Int.* **1992**, *5*, 38.

(29) Schlaad, H.; Schmitt, B.; Müller, A. H. E. *Angew. Chem., Int. Ed. Engl.* **1998**, *37*, 1389.

(30) Schmitt, B. Dissertation, University of Mainz, 1999.

(31) Hong, S. C.; Matyjaszewski, K.; Gottfried, A. E.; Brookhart, M. *Polymeric Materials: Science and Engineering* **2001**, *84*, *in press*.

(32) Fox, P. A.; Waymouth, R. M.; Hawker, C. J. *Polym. Prepr. (Am. Chem. Soc., Div. Polym. Chem.)* **1999**, *40*, 872.

(33) Stehling, U. M.; Malmstrom, E. E.; Waymouth, R. M.; Hawker, C. J. *Macromolecules* **1998**, *31*, 4396.

(34) Johnson, L. K.; Mecking, S.; Brookhart, M. *J. Am. Chem. Soc.* **1996**, *118*, 267.

(35) Von Werne, T.; Patten, T. E. *J. Am. Chem. Soc.* **1999**, *121*, 7409.

(36) Pyun, J.; Matyjaszewski, K.; Kowalewski, T.; Savin, D.; Patterson, G.; Kickelbick, G.; Huesing, N. *Polym. Prepr. (Am. Chem. Soc., Div. Polym. Chem.)* **2001**, *42*, 223.

(37) Matyjaszewski, K.; Miller, P. J.; Shukla, N.; Immaraporn, B.; Gelman, A.; Luokala, B. B.; Siclovan, T. M.; Kickelbick, G.; Vallant, T.; Hoffmann, H.; Pakula, T. *Macromolecules* **1999**, *32*, 8716.

(38) Husseman, M.; Malmstroem, E. E.; McNamara, M.; Mate, M.; Mecerreyes, D.; Benoit, D. G.; Hedrick, J. L.; Mansky, P.; Huang, E.; Russell, T. P.; Hawker, C. J. *Macromolecules* **1999**, *32*, 1424.

(39) Gaynor, S. G.; Matyjaszewski, K. *Controlled Radical Polymerizations*; American Chemical Society: Washington, DC, 1998; Vol. 685, p. 396.

(40) Fonagy, T.; Ivan, B.; Szesztay, M. *Macromol. Rapid Commun.* **1998**, *19*, 479.

(41) Hong, S. C.; Pakula, T.; Matyjaszewski, K. *Macromol. Chem. Phys.* **2001** *submitted*.

(42) Matyjaszewski, K.; Teodorescu, M.; Miller, P.; Peterson, M. L. *J. Polym. Sci., Part A: Polym. Chem.* **2000**, *38*, 2440.

(43) Simonazzi, T.; De Nicola, A.; Aglietto, M.; Ruggieri, G. *In Comprehensive Polymer Science, 1st Suppl.*; Pergamon Press: Oxford, **1992**, p. 133.

(44) Chung, T. C. *ACS Symp. Ser.* **2000**, *749*, 104.

(45) Percec, V.; Barboiu, B. *Macromolecules* **1995**, *28*, 7970.

(46) Matyjaszewski, K.; Beers, K. L.; Coca, S.; Gaynor, S. G.; Miller, P. J.; Paik, H.-j.; Teodorescu, M.

Polym. Prepr. (Am. Chem. Soc., Div. Polym. Chem.) **1999**, *40*, 95.

(47) Matyjaszewski, K.; Gaynor, S. G.; Coca, S. *PCT Int. Appl.* **1998**.

(48) Paik, H.-j.; Gaynor, S. G.; Matyjaszewski, K. *Macromol. Rapid Commun.* **1998**, *19*, 47.

(49) Nakagawa, Y.; Miller, P. J.; Matyjaszewski, K. *Polymer* **1998**, *39*, 5163.

(50) Beers, K. L.; Gaynor, S. G.; Matyjaszewski, K.; Sheiko, S. S.; Möller, M. *Macromolecules* **1998**, *31*, 9413.

(51) Beers, K. L.; Gaynor, S. G.; Matyjaszewski, K.; Sheiko, S. S.; Prokhorova, S. A.; Möller, M. *Polym. Prepr. (Am. Chem. Soc., Div. Polym. Chem.)* **1999**, *40*, 446.

(52) Börner, H. G.; Beers, K. L.; Matyjaszewski, K.; Sheiko, S. S.; Möller, M. *Macromolecules* **2001**, *in press*.

(53) Beers, K. L.; Matyjaszewski, K. *unpublished results*.

(54) Yamada, K.; Miyazaki, M.; Ohno, K.; Fukuda, T.; Minoda, M. *Macromolecules* **1999**, *32*, 290.

Living Radical and Cationic Polymerizations in Water and Organic Media

Mitsuo Sawamoto, Masami Kamigaito*

Department of Polymer Chemistry, Graduate School of Engineering,
Kyoto University, Kyoto 606-8501, Japan
[e-mail: sawamoto@star.polym.kyoto-u.ac.jp]

Summary: This lecture will focus two living polymerizations that can be carried out in water as well as in organic media. The first one is transition metal-catalyzed living radical polymerization, for which Ru(II) and other transition metals play a critical role to control the process; for aqueous systems, Ru(II) and Fe(II) half metallocene complexes are useful. The second is cationic polymerization with water-tolerant Lewis acids as catalysts, including rare earth triflates and boron trifluoride for selected monomers. Discussion will be directed to the design of initiating systems, search of versatile catalysts, and precision synthesis of new polymers.

Introduction

Despite the fact that both radical and cationic polymerizations have been difficult to control, primarily due to, respectively, bimolecular termination of free radical intermediates and chain transfer (β-proton elimination) of carbocationic growing species, we have shown that they can be finely controlled to give *living* radical and cationic polymerizations. Both systems can be initiated with halogen-containing initiators (R–X; X = halogen) coupled with appropriate catalysts (activators): transition metal complexes for radical living polymerization and selected Lewis acids (mostly metal halides) for cationic counterparts. In these living polymerizations, it is now widely accepted that the carbon–halogen terminals, derived from the initiators R–X, form the so-called "dormant" species which can be dissociated into radical or carbocationic growing species by the relevant catalyst. Importantly, the dissociation processes [—C–X + catalyst $\rightarrow$ —C* X(catalyst); C* = radical C$^\bullet$ or carbocation C$^+$] are reversible and in a rapid, dynamic equilibrium, so as to retain a relatively low concentration of the truly active intermediate (—C*) and thus to effectively minimize the side-reactions (termination or chain transfer).

This paper presents a brief overview on these two living polymerizations that we have recently developed, with some emphasis focused on the processes operable in water or similar aqueous media where neither traditional metal-catalyzed nor carbocationic polymerization has been perceived controllable and, above all, possible.

                     CCC 1022-1360/00/$ 17.50+.50/0

Transition Metal-Catalyzed Living Radical Polymerization

In 1994–95, we reported the first example of living radical polymerization mediated by a transition metal catalyst[1-3]. Figure 1 illustrates a typical example of such processes where a ruthenium(II) dichloride complex, in conjunction with a halogen-containing initiator, induces a living polymerization of methyl methacrylate (MMA) to give polymers of controlled molecular weights and very narrow molecular weight distributions (MWDs).

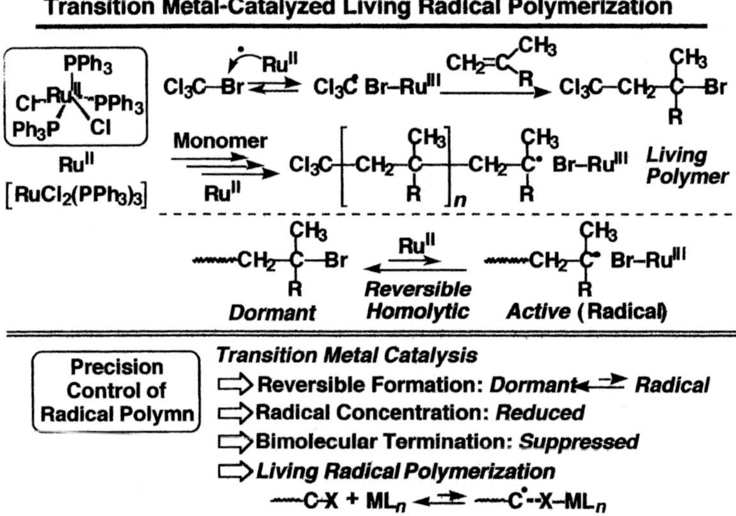

Figure 1. Transition metal-catalyzed living radical polymerization: A typical example and the general principles for precise control of radical polymerization[2,3].

As discussed above in a general form, the key to these living radical polymerizations is the reversible formation of radical species from the dormant carbon–halogen terminal assisted by the Ru(II) catalyst, which undergoes a one-electron oxidation into a Ru(III) species that, after some propagation step, regenerates the dormant end via a one-electron reduction. The intermediate radical concentration is thus kept much lower than in conventional free radical polymerizations, and thereby bimolecular termination is effectively suppressed relative to propagation.

(a) Scope

The following five years or so have witnessed a rapid, systematic, and worldwide development of the transition metal-catalyzed living radical polymerizations in terms of the scope of monomers, catalysts, and initiators, along with the related "precision" synthesis of finely controlled and designed polymers. For example, Figure 2 gives just a small list of catalysts, showing which catalyst is effective or applicable to which monomers of different classes (methacrylates, acrylates, styrenes, etc.). We are currently interested in "half-metallocene" complexes of Ru(II) and Fe(II), because, as shown in the table, they are more active and, equally important, more versatile (applicable to more monomers) than the now classical dichloride/phosphine counterparts[4]. The substituents for these metallocenes are typically indenyl (Ind), cyclopentadienyl (Cp), and pentamethylcyclopentadienyl (Cp*).

Metal-Catalyzed Living Radical Polymerization: Typical Catalysts vs. Monomers

Metal Catalyst \ Monomer	$CH_2=C(CH_3)C(=O)OCH_3$	$CH_2=CH$ (phenyl)	$CH_2=CH\,C(=O)OCH_3$	$CH_2=CH\,C(=O)N(CH_3)_2$
$RuCl_2(PPh_3)_3$	◯ (–Cl, –Br)	△ (–Br, –I)	△ (–Br, –I)	△ (–Br)
RuCl(Ind)(PPh$_3$)$_2$	◯ (–Cl)	◯ (–Br)	△ (–Cl, –Br)	✕
RuCl(Cp*)(PPh$_3$)$_2$	◯ (–Cl)	◯ (–Cl, –Br)	◯ (–Cl, –Br)	—
FeX(Cp*)(CO)$_2$	✕	◯ (–I)	◯ (–I)	△ (–I)

◯: Living / Narrow MWD;   △: Living / Broad MWD;   ✕: Not Living

Figure 2. The scope of transition metal-catalyzed living radical polymerization: Typical metal catalysts and monomers.

Another important development, in our view, is that some additives can accelerate these Ru(II)- and Fe(II)-catalyzed polymerizations, and these additives include metal alkoxides [Al(OiPr)$_3$, etc.] and amines [Bu$_2$NH, etc.]. In a typical example, the RuCl(Ind)-(PPh$_3$)$_2$ (Ind = indenyl; Ph = phenyl) catalyst can polymerize MMA in toluene at 80 °C in ca. five hours to nearly quantitative conversion, where polymers have controlled molecular weights and dispersity ratios below 1.15. On the other hand, FeX(Cp*)-(CO)$_2$ (X = halogen) seems effective for styrene, giving similar living polymers from iodide-type initiators.

(b) Living Random Copolymerization

This subject is interesting, because it integrates the advantage of living processes (molecular weight control etc.) with those of radical polymerizations over ionic counterparts that a wide variety of monomer pairs can readily be copolymerized. In fact, the transition metal catalysts discussed above are equally effective in such "living" random copolymerizations, and some examples are compiled in Figure 3, showing typical catalysts for the MMA–methyl acrylate pair. Again, the Cp* catalysts are versatile and effective, and can form living random copolymers of narrow MWDs and molecular weights exceeding 10^5.

Living Random Copolymerization: Typical Metal Catalysts vs. Monomers

Metal Catalyst \ Monomer	Homopolymerization		MMA/MA Random copolymer
	MMA $CH_2=C(CH_3)(C=O)OCH_3$	MA $CH_2=CH(C=O)OCH_3$	
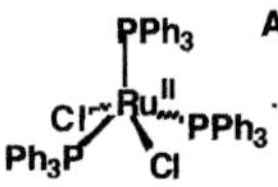 PPh_3 / $Al(Oi\text{-}Pr)_3$ — Cl–Ru^{II}–PPh_3, Ph_3P, Cl	◯ (–Cl, –Br)	△ (–Br, –I)	◯ (–Cl) ┊ ◎ (–Br)
Amine	◯ (–Cl, –Br)	✕	◯ (–Cl)
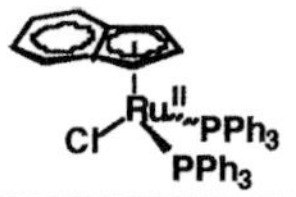 (indenyl) Cl–Ru^{II}–PPh_3, PPh_3	◯ (–Cl)	△ (–Cl, –Br)	◯ (–Cl)
 Br–Ni^{II}–$PnBu_3$, nBu_3P, Br	◯ (–Br)	△ (–Br)	◯ (–Br)
(Cp) Fe^{II}–CO, I, CO	✕	◯ (–I)	◯ (–I)
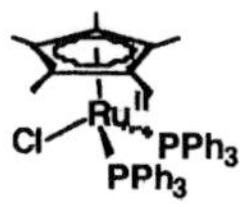 (Cp*) Cl–Ru^{II}–PPh_3, PPh_3	◯ (–Cl, –Br)	◯ (–Cl, –Br)	◎ (–Cl)

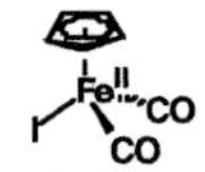

Homopolymn ◯: Living / Narrow MWD ($M_w/M_n = 1.1$–1.2)
△: Living / Broad MWD   ✕: Not Living

Copolymn ◎: Living / Narrow MWD ($M_w/M_n < 1.5$)
◯: Living / Broad MWD ($M_w/M_n > 1.5$)

Figure 3. Living radical copolymerization with transition metal catalysts: Typical catalysts and their applicability for the methyl methacrylate–methyl acrylate comonomer pair.

(c) Core-Functionalized Star Polymers

Along with the development of initiating systems and metal catalysts, we have been pursuing the precision synthesis of new polymers via our living radical polymerizations[3]. A recent example of interest is "core-functionalized" star polymers (Figure 4): Linear living polymers, obtained in a metal-mediated polymerization and kept unquenched, are treated with a small amount of selected divinyl monomers, which are attached to the polymer end as a short segment carrying unreacted vinyl functions in the pendant group. These vinyl moieties then react with the growing radical ends, initiating a "polymer linking" process, to give microgel cores to which a large number of the linear polymers are attached as "arms".

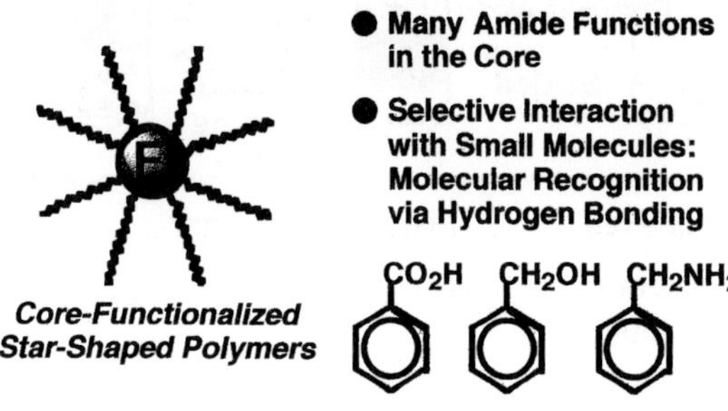

Figure 4. Core-functionalized star polymers by metal-mediated living radical polymerization and polymer linking with a bifunctional acrylamide.

The example shown in Figure 4 is specifically employs a series of bifunctional amide monomers [e.g., $CH_2=CH(CO)NHCH_2NH(CO)CH=CH_2$] as polymer-linking agents, and thus the resulting star polymers carry a large number (sometimes more than a few hundred per core) of amide functions in the microgel core, then coined core-functionalized.

Thanks to these concentrated and localized amide functions, these star polymers can interact with some polar organic compounds, such as benzoic acid, through hydrogen bonding with the amide group.

(d) Living Radical Polymerization in Water

In general, transition metal complexes are not stable in water and similar polar media, often loose their catalytic activity, but some class of Ru(II) compounds are known to be less oxophilic and thereby active as catalysts even under aqueous conditions. This is indeed the case for the Ru(II)-catalyzed living radical polymerization (Figure 5).

For example, the three catalysts shown in Figure 5 can be employed for MMA and other monomers that are polymerized efficiently into living polymers of controlled and sometimes fairly high molecular weights beyond 10^5. Typically, the monomer, an ini-

22

tiator, and a Ru(II) catalyst are mixed either in bulk monomer or in a small amount of
toluene, and the solution is vigorously mixed into an excess of water to form a suspension. On elevating temperature, living polymerization has been found to proceed in the
fine droplets, as in those in organic media.

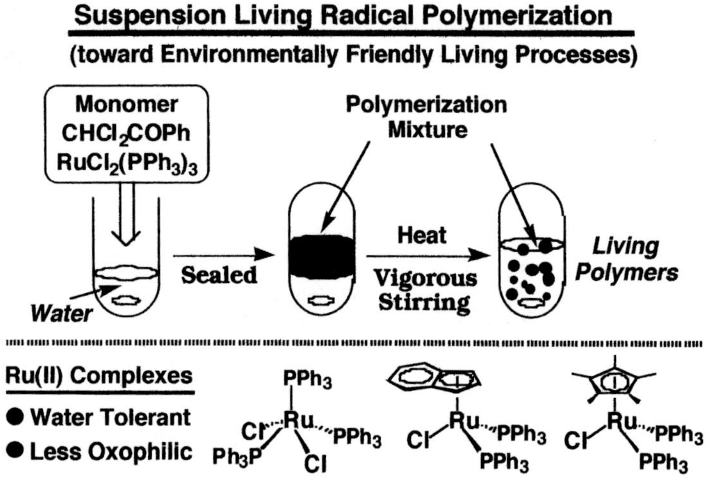

Figure 5. Suspension living radical polymerization in water: A schematic illustration.

Recent results in our laboratories also show that $FeXCp^*(CO)_2$ is similarly tolerant to
water and catalyzes living radical polymerizations of acrylates and styrene.

As readily expected, these living processes feasible in water would be of interest as
future production of polymers under environmentally benign conditions. No doubt, the
critical factor is to develop metal catalysts that withstand an excess of water, and
another aspect would be water-soluble metal catalysts that can medicate similar precision polymerizations in homogeneous aqueous systems for water-soluble monomers
such as acrylic acid and acrylamide.

Cationic Polymerization in Water or in Aqueous Solution

In contrast to the notion that cationic polymerization must be carried out under strin-
gently anhydrous conditions, we have recently found that sometimes it is not the case.
As summarized in Figure 6, the use of water-tolerant Lewis acids as catalysts permits us
to carry out cationic polymerizations either in water (suspension or emulsion systems)
or in polar solvents containing excess water.

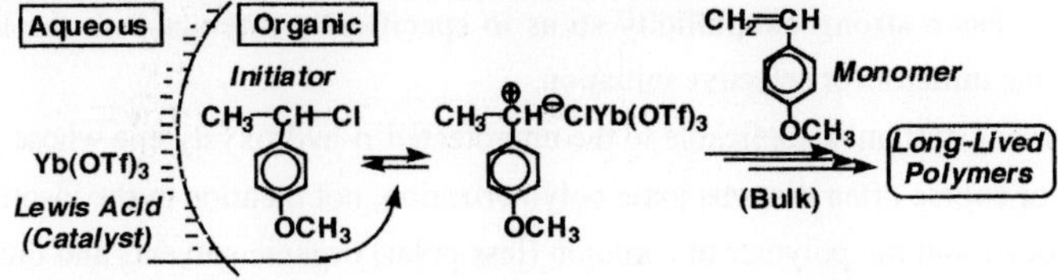

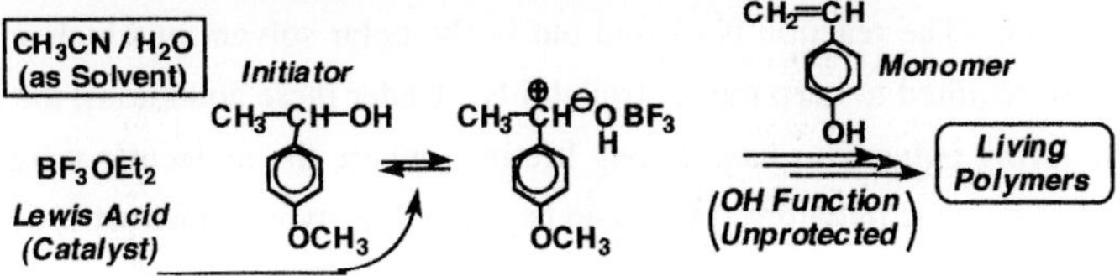

Figure 6. Cationic polymerizations in water or in aqueous solution with water-tolerant Lewis acid catalysts[5,6].

(a) Suspension and Emulsion Cationic Polymerization in Water into Long-Lived Polymers

One approach is to utilize rare-earth metal triflates (trifluromethanesulfonates) as catalysts; because the large ionic radii and the high coordination number of these metals, along with the strong electron-withdrawing nature of the triflate substituent, these salts are highly soluble in water and exhibit strong Lewis acidity therein (Figure 6A)[5]. A typical example is ytterbium triflate, $Yb(OTf)_3$ ($OTf = OSO_2CF_3$), which turned out to induce cationic polymerization of vinyl ethers and p-alkoxystyrenes when couple with a suitable halogen-containing initiator such as the hydrogen chloride adducts of the latter monomers; for these monomers, also, the resultant polymers are of a long lifetime and their molecular weights increase with conversion.

These polymerizations can be performed either in suspension (similar to the corresponding living radical systems) or in emulsion (with tetraalkylammonium salts as surfactants), and separate analysis shows that they in fact involve carbocationic species surviving in the organic phase exposed to bulk water. Copper and zinc triflates may also be used in place of the rare-earth salts, though less active.

(b) Cationic Polymerization with Water for (Unprotected) p-Hydroxystyrene

The second approach is for homogeneous polymerization of unprotected p-hydroxystyrene (p-vinylphenol) in acetonitrile solvent containing an excess of water[6]. The Lewis acid of choice herein is, rather surprisingly, boron trifluoride etherate (BF_3OEt_2), one of the most conventional and oft-employed Lewis acids for cationic polymerization, and it is also important to combine it with initiators where the hydroxyl group is the anionic leaving group for carbocation formation (Figure 6B). The boron compound apparently has a strong oxophilicity so as to specifically interact with the hydroxyl group in the initiator for selective initiation.

This initiating system is applicable to the unprotected p-hydroxystyrene whose phenolic group is of course often disturbs ionic polymerization, not mention to the insolubility of the monomer and the polymer in common (less polar) organic solvents and thus further complicate its direct polymerization. Nevertheless, the BF_3OEt_2-catalyzed process withstands the phenolic groups in the monomer and cleanly induces its cationic polymerization. The reaction is carried out in the polar solvent in which addition of water is also required to keep the controllability. Under these conditions, the products, similar to living polymers, have a long lifetime, where the molecular weight can be controlled. The same initiating system can be applied to styrene polymerization.

[1] T. Higashimura, M. Sawamoto, M. Kamigaito, Jpn. Patent Appl. 179850 (Aug. 1994); Kokai, 41117 (Feb. 13, 1996).
[2] (a) M. Kato, M. Kamigaito, M. Sawamoto, T. Higashimura, *Macromolecules* **1995**, *28*, 1721. (b) T. Ando, M. Kato, M. Kamigaito, M. Sawamoto, T. Higashimura, *Macromolecules* **1996**, *29*, 1070.
[3] For a recent review: M. Sawamoto, M. Kamigaito, *CHEMTECH* **1999**, *29*(6), 30.
[4] (a) H. Takahashi, M. Kamigaito, M. Sawamoto, *Macromolecules* **1999**, *32*, 3820. (b) Y. Kotani, M. Kamigaito, M. Sawamoto, *Macromolecules* **2000**, *33*, 3543.
[5] K. Satoh, M. Kamigaito, M. Sawamoto, *Macromolecules* **1999**, *32*, 3827; *Macromolecules* **2000**, *33*, 4660; *Macromolecules* **2000**, *33*, 5836; *J. Polym. Sci., Part A: Polym. Chem.* **2000**, *38*, 2728.
[6] K. Satoh, M. Kamigaito, M. Sawamoto, *Macromolecules* **2000**, *33*, 5405; *Macromolecules* **2000**, *33*, 5830; *Macromolecules* **2001**, *34*, in press.

Free-Radical Synthesis of Block Copolymers on an Industrial Scale

Benedikt Raether[1], *Oskar Nuyken*[2], *Philipp Wieland*[2], *Wolfgang Bremser*[3]

[1]BASF Future Business GmbH, 67056 Ludwigshafen, Germany
[2]Technische Universität München, Lehrstuhl für Makromolekulare Stoffe, Lichtenbergstraße 4, 85747 Garching, Germany
[3]BASF Coatings, Glasuritstraße 1, 48165 Münster, Germany

Summary: Controlled free-radical polymerization has been monitored with great interest in recent years since it offers an opportunity to combine the advantages of conventional free-radical polymerization with those of living ionic polymerization. We present the 1,1-diphenylethene (DPE) method which enables us to produce block copolymers on an industrial scale by a free-radical mechanism.

This DPE process enables industrially relevant monomers, such as styrene, methacrylates, acrylates, methacrylic acid, acrylic acid and N-vinyl compounds, to be converted into block copolymers. The synthesis can be carried out in organic solvents, without solvents or in water.

We have been able to demonstrate, that the addition of 1,1-diphenylethylene to a normal free-radical polymerization results in polymers whose molar mass, after a short uncontrolled phase, increases in a linear manner with conversion. The amount of 1,1-diphenylethylene added also determines the order of magnitude of the final molar mass. It was also possible to employ the polymers isolated during this polymerization as initiators for the polymerization of a further monomer, resulting in the formation of block copolymers.

With possibly somewhat reduced claims on the perfection of the structures, a wide variety of possibilities arise with the known advantages of free-radical polymerization.

The one-pot synthesis is carried out by simple successive addition of the desired monomers and has already been used successfully on an industrially relevant scale.

Introduction

Until recently polymerizations with a high degree of structure control were only possible by means of relatively complex methods, such as anionic, cationic or group-transfer polymerization. But these methods have the disadvantage that the high structure control is balanced by a limited choice of monomers and the high sensitivity of the reaction to impurities. Conventional free-radical polymerization on the other hand is a relatively insensitive process with high tolerance to functional groups, but without the possibility of structure control.

    CCC 1022-1360/00/$ 17.50+.50/0

Controlled free-radical polymerization combines the advantages of ionic and free-radical polymerization: a wide range of monomers with a high degree of structure control. It is ideally insensitive and can be carried out in existing plants.

The high structure control facilitated by controlled free-radical polymerization is of course not an end in itself, but has three industrial benefits:

1. The synthesis of gradient, block or comb polymers instead of random copolymers enables the production of materials with new or improved properties which are suitable, for example, as dispersants and emulsifiers, as thickeners, impact modifiers, phase promoters or adhesives. The low polydispersity in these cases is of minor concern.
2. On the other hand, low polydispersity of block copolymers or even homopolymers may be important for applications in pharmacology or powder coatings.
3. In functional polymers, expensive monomers which are the actual active component can be partially replaced by a less expensive monomer without impairing the effectiveness of the product.

We would like to pick out one of these potential areas of application of block copolymers, namely dispersants, in order to underscore the economic importance.
In 1995 600,000 metric tons of dispersants were produced in the USA, somewhat more than half of which were polymers. In 2000 it was almost 700,000 metric tons of dispersants. The value of polymeric dispersants produced in the USA in 1995 was almost \$500 million, in 2000 it was more than \$600 million.
Extrapolated to world production of polymers, this means that polymeric dispersants with a value of around \$2.1 billion were produced worldwide in the year 2000.

The first steps toward controlled free-radical polymerization were taken in the 1970's. Borsig and Braun polymerized MMA with diaryl- and triaryl-containing protecting groups.[1] They observed an increase in the molecular weight with conversion and the formation of block copolymers. Minoura polymerized MMA in the presence of chromium(II)acetate and again observed an increase in the molecular weight with conversion and the formation of block copolymers.[2]
In the 1980's, Otsu used dithiocarbamates as iniferters in the polymerization. He derived the term iniferters from the term inifer used in cationic polymerization. He also

used the term "living free-radical polymerization" for the first time.[3] The first controlled free-radical polymerization with a stable free radical appeared in the patent literature: Rizzardo and Solomon reported polymerization with an AIBN/TEMPO adduct.[4] However, the implications of this patent were not recognized and it was forgotten. Rizzardo did not follow up this concept and used a different mechanism: the RAFT mechanism.[5]

In the 1990's, eight years after the appearance of the Rizzardo and Solomon patent, controlled free-radical polymerization with stable free radicals was rediscovered by Georges. He published the controlled free-radical polymerization of styrene with TEMPO.[6] Matyjaszewski investigated the degenerative transfer of alkyl iodides for controlled free-radical polymerization.[7]

Also in the 1990s, Sawamoto and Matyjaszewski and other authors reported the controlled free-radical polymerization by ATRP with different metals, such as copper, ruthenium, iron, nickel and rhodium.[8-15]

The DPE Method

We have found that 1,1-diphenylethene is an additive, which also enables control of free-radical polymerization.[16-19] In a typical synthesis of the first block, the monomer, initiator and 1,1-diphenylethene are heated at 60 to 110°C for 2 to 12 hours. A second monomer is then added, and the mixture is heated at 60 to 110°C for a further 6 to 12 hours. Depending on the choice of monomer, this gives hydrophobic, amphiphilic or water-soluble block copolymers in a one-pot synthesis.

The DPE method is surprisingly flexible with respect to the choice of monomers and the solvent. The method works with all common monomers, which are capable of free-radical homopolymerization or copolymerization, even N-vinyl compounds and vinyl acetate. The polymerization can be carried out in organic solvents, in water or without a solvent. The reaction conditions are the usual ones for conventional free-radical polymerization, which is very beneficial for industrial implementation of the DPE method. The products are colorless and odorless with molecular weights from 5000 to 500,000 g/mol, and have typical block copolymer behavior.

Using methyl methacrylate and n-butyl acrylate, we have produced block copolymers with various block length ratios and molecular weights of up to 500,000 g/mol. For example, materials are obtained which are tacky in the case of long n-butyl acrylate blocks and highly elastic in the case of long methyl methacrylate blocks.

Methyl methacrylate and styrene can be converted into block copolymers by the DPE method. The right curve in the GPC in figure 1 shows the first block of homomethyl methacrylate, and the left curve shows the block copolymer after reaction of the first block with styrene. It is clear from the curves that the polydispersity is higher than in other methods of controlled free-radical polymerization.

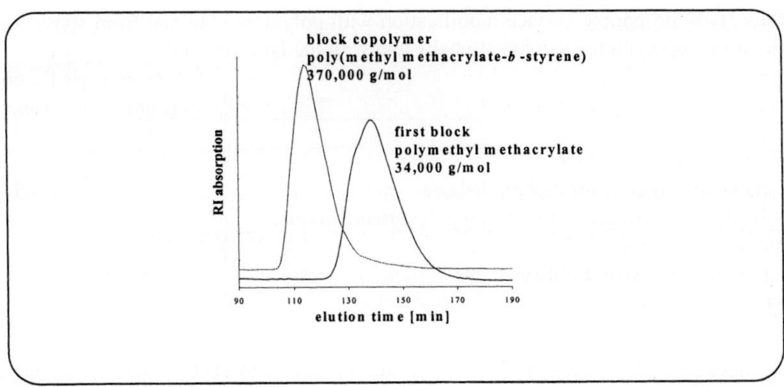

Figure 1: GPC curves of poly(methyl methacrylate), M_w = 34,000 g/mol, and the block copolymer poly(methyl methacrylate-*b*-styrene), M_w = 370,000 g/mol, obtained from it by the DPE method.

The DPE method also enables the production of block copolymers from styrene and maleic anhydride (MSA) (Figure 2). The conventional free-radical copolymerization of styrene and maleic anhydride gives a mixture of alternating maleic anhydride/styrene copolymer and homopolystyrene if styrene is present in excess. In the presence of 1,1-diphenylethene, however, the termination reaction can be suppressed and a pure styrene block is polymerized onto the first block of alternating styrene and maleic anhydride. This enables an amphiphilic block copolymer with novel interesting properties to be obtained from inexpensive monomers in a one-step one-pot synthesis.

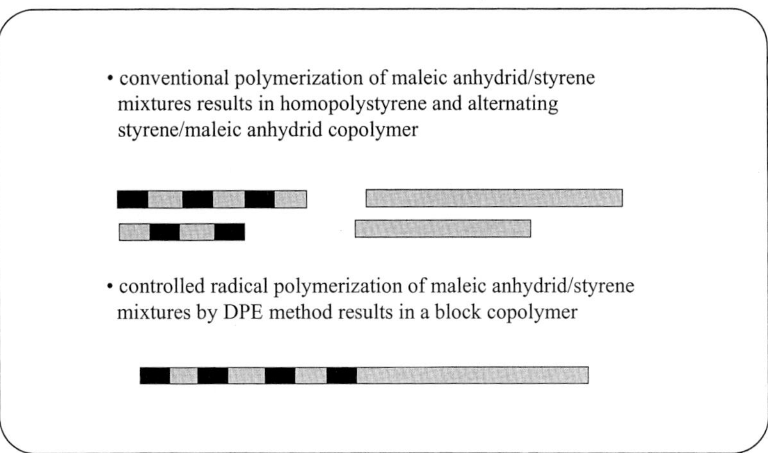

Figure 2: Styrene/MSA block copolymers by the DPE method.

Table 1: Hydrophobic surface modification with polymers obtained from styrene/MSA mixtures, copolymerized conventionally and with the DPE method.

	contact angle [$^{\circ}$] on	
	glass	paper
uncoated	65	12
styrene/maleic anhydride, copolymerized **without** diphenylethene poly(styrene-*alt*-maleic anhydride)/polystyrene mixture	106	65
styrene/maleic anhydride, copolymerized **with** diphenylethene poly((styrene-*alt*-maleic anhydride)-*b*-styrene)	118	110

We have reacted a styrene/maleic anhydride mixture with DPE to give an amphiphilic block copolymer and polymerized a styrene/maleic anhydride mixture with the same monomer ratio by a conventional method without DPE (Table 1). In a very simple experiment glass and paper were coated with these products and the contact angle of the surfaces was measured. The expected hydrophobicizing effect is evident in all cases. However, the effect is much stronger in the case of the block copolymer and the difference is greater the more polar the coated surface. The polar block interacts with the polar surface, causing the hydrophobic polystyrene blocks of the block copolymer to point away from the coated surface like a brush. If the polar and non-polar blocks are not chemically bonded, but are merely in the form of a physical mixture, as in the case of the styrene/maleic anhydride mixture polymerized without DPE, this self-orientation cannot occur, as impressively confirmed by the experiment.

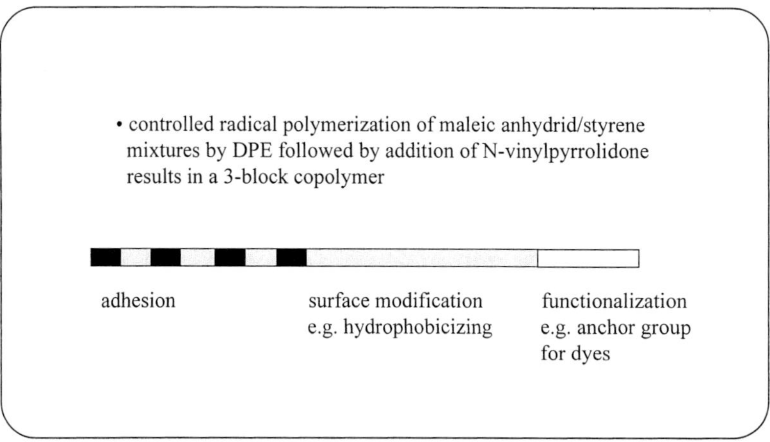

Figure 3: Poly((styrene-*alt*-MSA)-*b*-styrene-*b*-N-vinylpyrrolidone) by the DPE method.

The DPE method also enables a further block, in this case a polyvinylpyrrolidone block, to be polymerized onto a styrene-maleic anhydride block copolymer (Figure 3). Each block in this 3-block copolymer does a particular job: the block comprising alternating styrene and maleic anhydride causes adhesion through interaction with a polar surface, the styrene block renders the surface hydrophobic, and the polyvinylpyrrolidone block, owing to its complexing ability, is particularly suitable as anchor group, for example for dye molecules.

We have been able to produce other very interesting block copolymers using the DPE method. N-vinylpyrrolidone and styrene give amphiphilic block copolymers in various block length ratios. Depending on the block length ratio, the products are colloidally soluble, dispersible or only swellable in water. They have a strong complexing tendency, for example for oil and iodine, and the thickening effect which is typical of block copolymers. The electron photomicrograph (Figure 4) shows the continuous black styrene phase and the discontinuous pale N-vinylpyrrolidone phase. The N-vinylpyrrolidone/styrene block length ratio is 30/70.

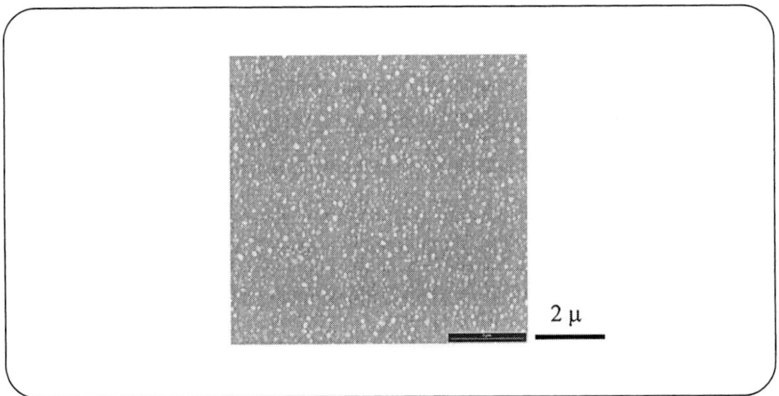

Figure 4: Electron photomicrograph of poly(styrene-*b*-N-vinylpyrrolidone) block copolymer obtained by the DPE method. The N-vinylpyrrolidone/styrene block length ratio is 30/70. The continuous black phase shows styrene and the discontinuous pale phase N-vinylpyrrolidone.

The DPE method can also be used to produce block copolymers from ethene and styrene. To this end, a DPE-containing polystyrene was reacted with ethene at 1300 bar and 250°C in an industrial plant, and a block copolymer was obtained in a exothermic reaction. Block copolymers of this type may be of great economic importance.

Demonstration of Block Structure

We have used various methods to demonstrate the block structure of the copolymers obtained by the DPE method.

Strong indications of the formation of block copolymers were firstly the observed monomer conversion and an increase in the molecular weight in the second step. The solubility behavior of amphiphilic block copolymers is characteristic and was also found in the copolymers produced by the DPE method: the addition of organic solvents gives opaque solutions or dispersions which suddenly became clear on addition of, for example, small amounts of water. GPC-IR coupling confirms the presence of block copolymers from the parallel intensities of the absorption in the GPC and in the IR at a wavenumber that is characteristic of one of the monomers. Pyrolysis GC-MS coupling clearly showed the presence of fragments consisting of two different chemically linked monomer types. We have also carried out extraction experiments and recorded electron photomicrographs, which we would like to explain in somewhat greater detail.

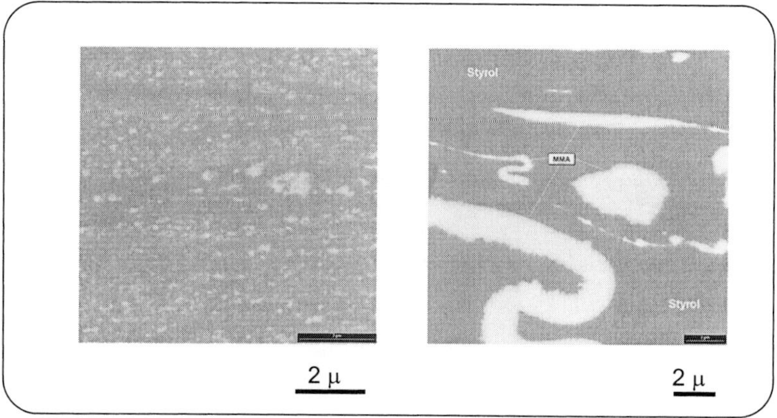

Figure 5: Electron microscopy picture of block copolymer of methyl methacrylate and styrene obtained by the DPE method (left) and of a mixture of polystyrene and polymethyl methacrylate (right). The molecular weights and mixing ratios are comparable.

In order to demonstrate that block copolymers really can be obtained by the DPE method, electron photomicrographs were recorded (Figure 5). On the left-hand side one can see the electron photomicrograph of a block copolymer of methyl methacrylate and styrene produced by the DPE method. On the right-hand side is the electron photomicrograph of a physical mixture of homopolystyrene and homopolymethyl methacrylate. The molecular weights and mixing ratios are comparable. Owing to incompatibility, the two polymers separate in the right-hand picture and form large domains. The styrene phase appears black and the methyl methacrylate phase white. In the left-hand picture, by contrast, there appears to be a separation on a smale scale in spite of the incompatibility as it is typical for block copolymers. The different polymer blocks must be chemically linked, i.e. a block copolymer is present.

Further evidence of the block structure in the copolymers produced by the DPE method was provided by extraction experiments (Figure 6). To this end, a block copolymer of methyl methacrylate and styrene produced by the DPE method was extracted with cyclohexane in a Soxhlet apparatus. Cyclohexane is a solvent for homopolystyrene, but not for homopolymethyl methacrylate. After the extraction, an IR spectrum of the residue was recorded.

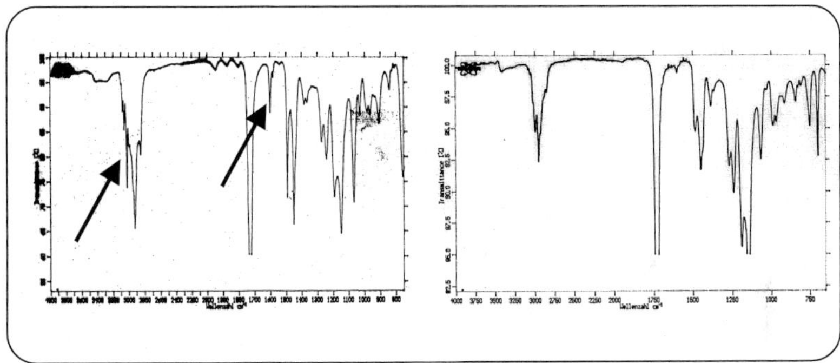

Figure 6: IR-Spectra of block copolymer of methyl methacrylate and styrene obtained by the DPE method (left) and of a mixture of polystyrene and polymethyl methacrylate (right) after extraction with hot cyclohexane. The molecular weights and mixing ratios are comparable. Arrows indicate styrene signals.

The IR spectrum on the left-hand side clearly shows the styrene signals (see arrows) in spite of the preceding cyclohexane extraction. The right-hand picture shows the IR spectrum of the residue of the physical mixture of homopolystyrene and homopolymethyl methacrylate which has been subjected to the same extraction with cyclohexane in a Soxhlet apparatus. The molecular weights and mixing ratios are comparable. The IR spectrum no longer contains a signal for styrene. This means that a block copolymer really must be present in the case of the copolymer produced by the DPE method.

We have recorded an electron photomicrograph of a 3-block copolymer of n-butyl acrylate, styrene and n-butyl acrylate in order to demonstrate the presence of a block copolymer (Figure 7). The black, continuous phase shows the styrene phase, and the white, discontinuous phase shows the n-butyl acrylate phase. This is a block copolymer with a molecular weight of around 370,000 g/mol and a polydispersity of 2.6. As already mentioned, the block copolymers synthesized by the DPE method are not distinguished by particularly low polydispersity. This also means that although the electron photomicrographs show the small domains that are typical of block copolymers, they do not show the usual high degree of order.

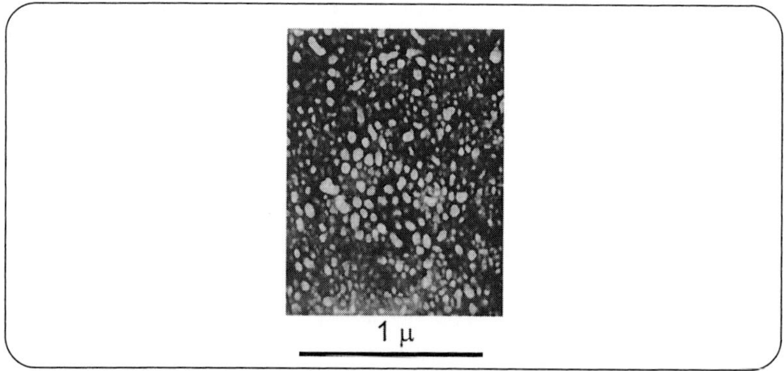

Figure 7: Electron microscopy picture of a A-B-A 3-block copolymer of n-butyl acrylate, styrene and n-butyl acrylate (M_w = 366,000 g/mol, M_w/M_n = 2.6), stained with RuO_4.

Block Copolymers from Anionically Polymerized DPE Copolymers

"Super Polystyrene", a former research product of BASF, is a copolymer of styrene and 1,1-diphenylethene (Figure 8), which has a higher glass transition temperature than pure polystyrene. The glass transition temperature increases with the DPE content. It is synthesized by anionic copolymerization.

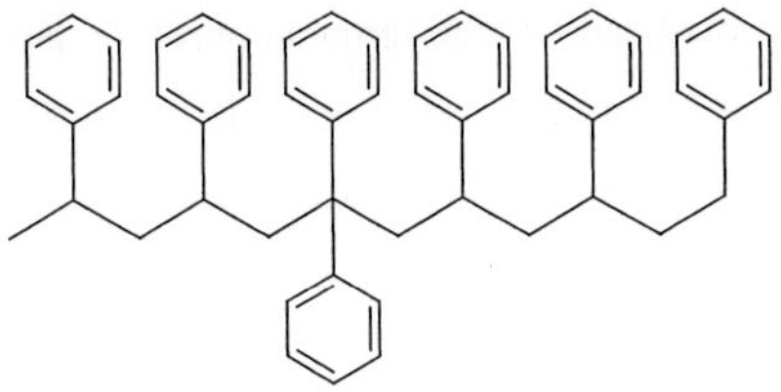

Figure 8: Structure of anionically polymerized styrene-DPE copolymer.

Against the background that polymers containing DPE can be reacted with a further monomer by a free-radical mechanism to give block copolymers, it was interesting whether it is possible for a "super polystyrene" of this type to be reacted with a monomer which is capable of free-radical polymerization to give block copolymers. In fact, by simply warming a "super polystyrene" with a molecular weight of 180,000 g/mol with a monomer that is capable of free-radical polymerization, in this case methyl methacrylate, we have obtained a corresponding block copolymer with a molecular weight of 215,000 g/mol. For it a copolymer, containing 90 % styrene and 10 % 1,1-diphenylethene was dissolved in methylmethacrylate and heated up to 105 $^{\circ}$C for 8 h.

The DPE Mechanism

The mechanism of the DPE method is not yet clear. We have indications which make various mechanisms appear plausible. It is also possible for more than one mechanism to take place, perhaps depending on the monomer.

The mechanism shown in figure 9 is, that the growing polymer chain can be blocked reversibly with a DPE molecule, and so enables the successive incorporation of various monomers. An irreversible head-to-head termination of the polymeric free radicals can take place as a side reaction. It is also possible for the diphenylethyl-terminated polymeric free radicals to combine. However, this should result in a highly substituted ethane bond, which, as it is thermally labile, can be broken again and finally can release the polymeric free radical again.

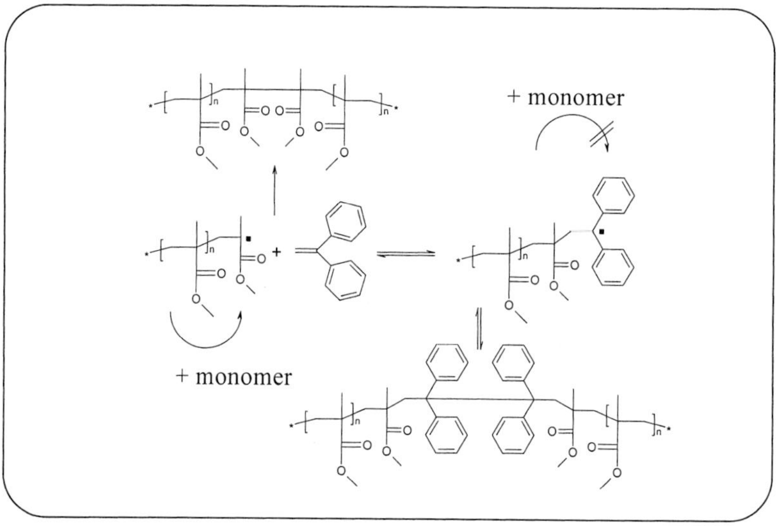

Figure 9: DPE mechanism 1.

Another possible mechanism is for 1,1-diphenylethene to be incorporated into the polymer chain or to retard the growth of a polymer chain and finally to combine with a further growing polymer chain in a head-to-head termination reaction (Figure 10). Both cases result in a polymer structure which includes a 1,1-diphenylethyl unit as a sort of predetermined breaking point. In a second step (Figure 11), the diphenylethyl unit in the para-position is attacked by a free radical and finally reversibly liberates a polymeric free radical, which is able to adduct a further monomer and so results in block copolymers. The driving force for this reaction could be the formation of a quinoid structure with the associated extension of the delocalized π system of 6 from 12 electrons. A point in favor of this assumption is that the block copolymer formation can be considerably accelerated by the continuous addition of small amounts of initiator.

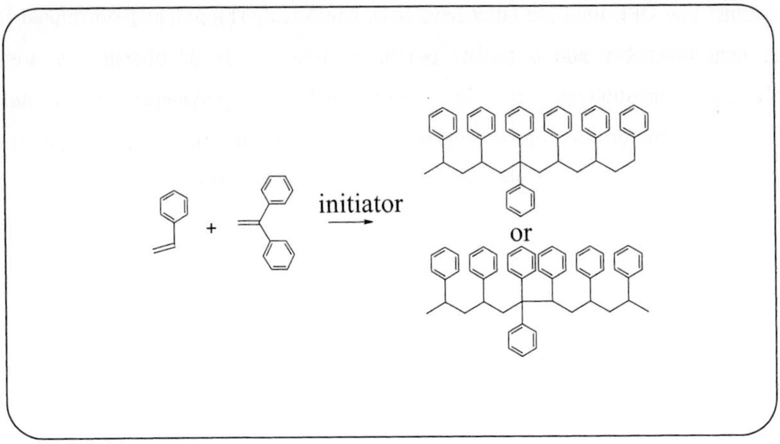

Figure 10: DPE mechanism 2, first step.

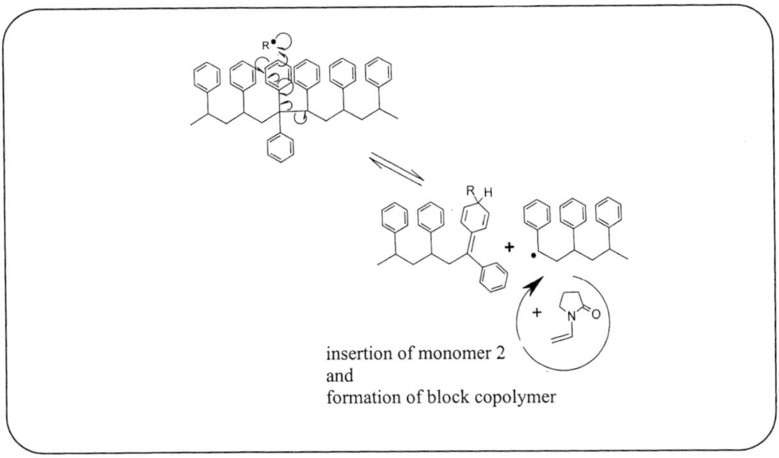

Figure 11: DPE mechanism 2, second step.

38

A further mechanism which appears plausible is based on addition of 1,1-diphenylethene onto the growing polymer chain during polymerization of the first block (Figure 12). After a temporary retardation of the polymerization process, a polymer-substituted diphenylethene and a new polymeric free radical are formed in a transfer reaction. The polymeric free radicals formed in this way react with further monomer and DPE until the latter have been consumed. The polymer-substituted DPE can then reversibly add a further polymeric free radical or liberate the original polymeric substituents in the form of a polymeric free radical (Figure 13). One or other of the polymeric free radicals are liberated alternately in this equilibrium reaction and is able to adduct a further monomer. At the same time, the concentration of free polymeric free radicals is suppressed to the extent that the combination reaction plays only a secondary role, but the desired polymerization reaction continues to occur.

In a side reaction, the diphenylethyl free radical substituted by two polymer chains can be converted into a diphenylethene derivative substituted by two polymer chains, with elimination or transfer of a hydrogen free radical, enabling the reaction to gradually subside. A further point in favor of this assumption is that the block copolymer formation can be considerably accelerated by the continuous addition of small amounts of initiator. In addition, the corresponding double bonds are detected in the ^{13}C-NMR.

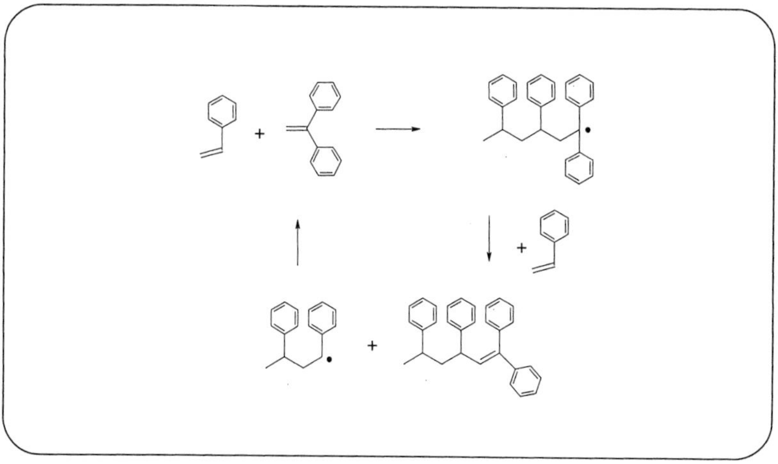

Figure 12: DPE mechanism 3, first step

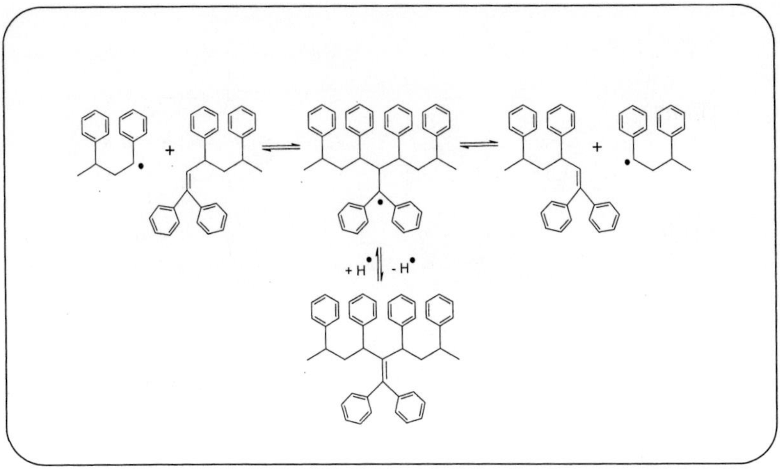

Figure 13: DPE mechanism 3, second step

The DPE method appears to be more versatile than other methods with respect to the choice of monomer. The fields highlighted in figure 14 show example structures which would reasonably be expected in the controlled free-radical polymerization of, for example, N-vinylpyrrolidone using the TEMPO or ATRP method. Either mixed N,O-acetals or geminal N-halogen compounds are formed here. However, both structures are very sensitive to elimination reactions and other side reactions at the reaction temperatures necessary for the respective reaction mechanism. The DPE method circumvents these problems since there are no heteroatoms to form structures which would facilitate the side reactions mentioned. This applies to all possible DPE mechanisms mentioned.

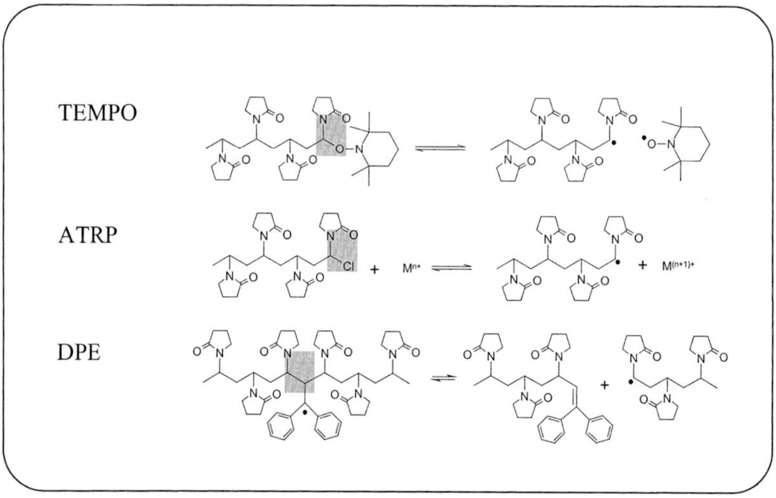

Figure 14: Versatility of the DPE-method

Conclusion

The DPE method enables the free-radical synthesis of block copolymers from all common free-radical homopolymerizable or copolymerizable monomers in organic solvents, water or without a solvent.

We have used the method to produce new materials, for example poly(N-vinylpyrrolidone-b-styrene), poly(N-vinylpyrrolidone-b-vinylacetate) and poly(styrene-b-ethene). In addition, the products are colorless, odorless and free from toxic additives. This is particularly important with regard to commercial applicability. The method is not suitable for the synthesis of polymers of low polydispersity. M_w/M_n is frequently between 2 and 3 and the materials produced in this way may contain a certain portion of homopolymer. However, both are negligible for many industrial applications. The DPE method does not compete with the known methods for controlled free-radical polymerization, but instead is intended to supplement the range of tools available to the polymer chemist. It is indisputable that all methods have their advantages and disadvantages. The mechanism is still unclear. We have indications which make one or other mechanism appear plausible, but as yet not all observations are consistent with a single mechanism alone. The most important factor is, however, that the DPE method enables us to produce block copolymers on an industrial scale, i.e. by the ton.

[1] E. Borsig, M. Lazar, M. Capla, S. Florian, *Angew. Makromol. Chem.* **1969**, *9*, 89
[2] M. Lee, Y. Minoura, *Chem. Soc. Faraday Trans. 1* **1978**, *74*, 1726
[3] T. Otsu, M. Yoshida, *Makromol. Chem. Rapid Commun.* **1982**, *3*, 127
[4] D. H. Solomon, E. Rizzardo, P. Cacioli, *U.S.Pat 4,581,429*; 1986
[5] E. Rizzardo, G.F. Meijs, S.H. Tang, *Macromol. Symp.* **1995**, *98*, 101
[6] M.K. Georges, R.P.N. Veregin, P.M. Kazmaier, G.K. Hamer, *Macromolecules* **1993**, *26*, 2987
[7] K. Matyjaszewski, S. Gaynor, J.S.Wang, *Macromolecules* **1995**, *28*, 2093
[8] J.S. Wang, K. Matyjaszewski, *J. Am. Chem. Soc.* **1995**, *117*, 5614
[9] M. Kato, M. Kamigaito, M. Sawamoto, T. Higashimura, *Macromolecules* **1995**, *28*, 1721
[10] J.S. Wang, K. Matyjaszewski, *Macromolecules* **1995**, 28, 7901
[11] V. Percec, B. Barboiu, A. Neumann, J.C. Ronda, M. Zhao, *Macromolecules* **1996**, *29*, 3665
[12] M. Wie, J. Xia, K. Matyjaszewski, *Polym. Prepr. (Am. Chem. Soc., Div. Polym. Chem.)* **1997**, *38(2)*, 231
[13] M. Wie, J. Xia, K. Matyjaszewski, *Polym. Prepr. (Am. Chem. Soc., Div. Polym. Chem.)* **1997**, *38(2)*, 233
[14] C. Granel, P. Dubois, R. Jerome, P. Teyssie, *Macromolecules* **1996**, *29*, 8576
[15] H. Uegaki, Y. Kotani, M. Kamigaito, M. Sawamoto, *Macromolecules* **1997**, *30*, 2249
[16] P. C. Wieland, B. Raether, O. Nuyken, *Makromol. Chem. Rapid Commun.* **2001**, *22(9)*, 700
[17] WO 00/37507
[18] WO 00/39169
[19] WO 01/44327

Contribution of Unsymmetrical Difunctional Initiators/Monomers to the Macromolecular Engineering

Robert Jérôme

Center for Education and Research on Macromolecules (CERM), University of Liège, Sart-Tilman, B6a, 4000 Liège, BELGIUM

Introduction

The emerging technologies (e.g. in optics, microelectronics, medicine…) require the availability of synthetic polymers with continuously more sophisticated properties and performances. The best way for the polymer chemist to face this challenge is to tailor the molecular structure of the chains. Nowadays, the progress in the living/controlled polymerization mechanisms is such that the so-called macromolecular engineering is a vivid reality.

Historically, the chemist created new polymers by combining two or more monomers in the same chain, so leading to (more or less) statistical copolymers. With the discovery of living polyadditions, the comonomers could be polymerized in a sequential way, and the advent of microphase separated block copolymers was a revolution in polymeric materials. The range of these block copolymers was increased by using two types of initiators also in a sequential way.[1] For instance, living polyanionic chains were reacted with an electrophile, such that a radical initiator or a precursor of a Ziegler-Natta catalyst is attached as an end-group and can initiate the polymerization of families of monomers that could not be polymerized by anions. So, combinations of single initiators and single monomers (one type of polymerizable moiety) was the strategy originally used in macromolecular engineering.

Recently, a new trend emerged which consists in associating in the same molecule, either two different initiators, or two different polymerizable moieties, or, more frequently, one monomer and one initiator. These dual molecules offer new prospects not only in the search for new polymers but also in the implementation of the synthesis of multicomponent materials. These three typical strategies will be illustrated in this paper in a non exhaustive way, mainly on the basis of experiments carried out by collaborators of the author.

Dual Monomer/Initiator Molecules

Any unsaturated and cyclic monomer substituted by a hydroxyl group (protected, if required) can be polymerized by polyaddition, and the pendant hydroxyl groups (after deprotection, if necessary) of the chains are potential initiators for the ring-opening polymerization (ROP) of lactones and lactides.

Although 2-hydroxyethylmethacrylate (HEMA) is a monomer known for a long time, its use as a dual monomer/initiator compound was not widespread for a long time. HEMA is actually a precursor of graft copolymers prepared according to the so-called "grafting-from" technique. Scheme 1 illustrates the living anionic copolymerization of presilylated HEMA with methyl methacrylate (MMA), followed by deprotection of the hydroxyl groups and their conversion into dialkylaluminum alkoxides, which are commonly used to initiate the controlled ROP of lactones and lactides.[2]

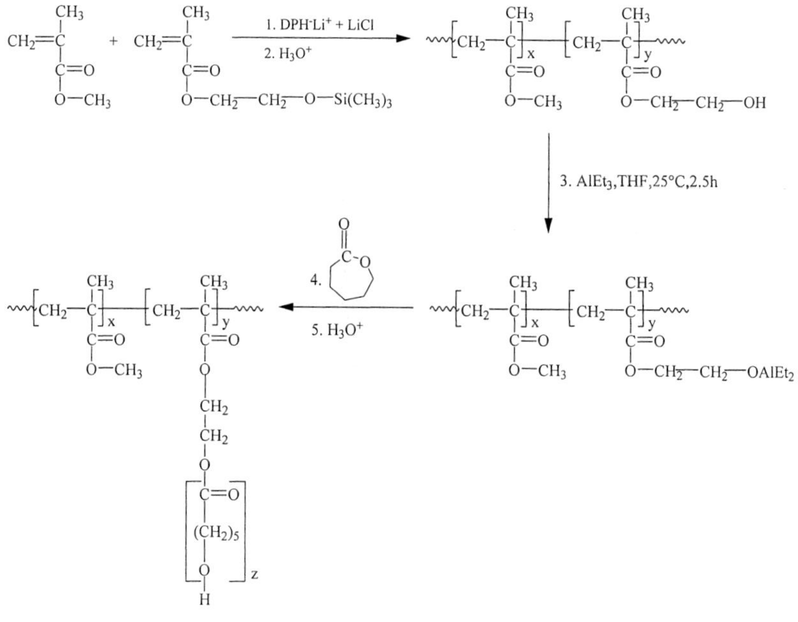

Scheme 1. HEMA : Key intermediate in the synthesis of PMMA-graft-poly(ε-caprolactone) copolymers

The anionic copolymerization was initiated by diphenylhexyllithium (DPH$^-$Li$^+$) in THF, at $-78°C$, in the presence of LiCl (10 equiv.) as a ligand. It was complete within a few hours. The livingness of the ligated polymerization allowed to control the molecular weight of the copolymer (by the comonomer/initiator molar ratio) and the grafting density (by the comonomer feed) and to trigger low polydispersity (1.05). The protecting trimethylsiloxy groups were stable towards the anions at $-78°C$ and quantitatively deprotected at $25°C$. The conversion of the hydroxyl groups into Al alkoxides was quantitative by reaction with a twofold molar excess of triethyl Al. Graft copolymers with M_n up to 2.10^5 g $\cdot$ mol^{-1} were prepared, that contained up to 25 grafts per molecule, M_n of the grafts being in the range from a few hundreds up to 20,000 g $\cdot$ mol^{-1}. The polydispersity was usually lower than 1.2. This system is very flexible because MMA can be substituted by other alkylmethacrylates and ε-caprolactone by lactides, δ-valerolactone and, better, by protected γ-hydroxyl-ε-caprolactone, which opens the way to more complex macromolecular architectures.[2] It must also be kept in mind that the hydroxyl groups of either the poly(MMA-co-HEMA) copolymers or the polyester grafts (end-group or comonomer units) are reactive in polycondensation reactions, e.g. in the sol-gel process, with the purpose to produce organic-inorganic hybrid materials.[3] HEMA is then rather a dual monomer than a dual monomer/initiator compound. This remark is valid to other hydroxyl containing monomers. Last but not least, novel palm-tree like architectures can be designed by using the poly(MMA-co-HEMA) macroinitiator as one constituent of block copolymers,[2] as schematized below.

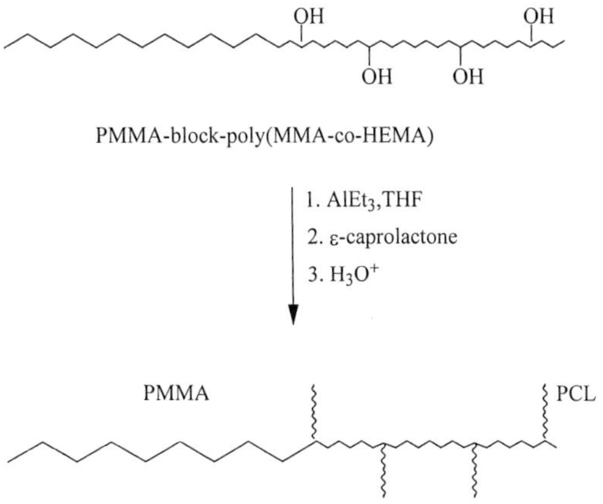

PMMA-block-poly(MMA-co-HEMA)

1. AlEt$_3$,THF
2. ε-caprolactone
3. H$_3$O$^+$

Scheme 2. Synthesis of "palm-tree" shaped block copolymers

Finally, controlled radical polymerization (CRP) could be carried out instead of living anionic polymerization in the first step of scheme 1, such that combinations of anionic/ROP and CRP/ROP can be envisioned.

As an answer to the urgent need to functionalize biodegradable and biocompatible aliphatic polyesters, γ-functional ε-caprolactones have been recently synthesized, particularly monomers γ-substituted by a protected hydroxyl group.[4,5] The acetal group attached to ε-caprolactone in γ-position in 1,4,8-trioxaspiro-[4,6]-9-undecanone (TOSUO) can be quantitatively derivatized into a hydroxyl group when part of a poly(ε-CL-co-polyTOSUO) copolymer. The 2-step derivatization is shown in scheme 3. Better, γ-(t-butyldimethylsilyloxy) ε-caprolactone (SCL) or triethylsilyloxy ε-caprolactone (TeSCL) can be copolymerized with ε-CL in a controlled way, and the one-step deprotection of the (Te)SCL comonomer units is straightforward. These deprotection reactions leave the polyester chains unaffected.

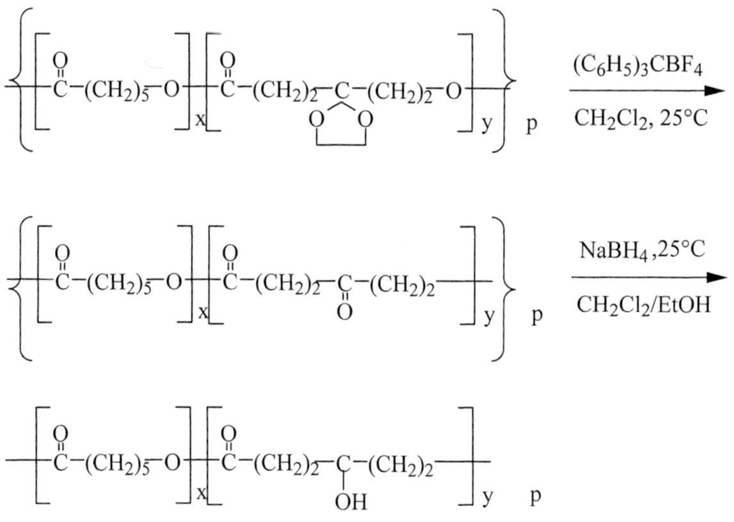

Scheme 3. Deprotection of γ-acetal pendant groups

The reaction of the hydroxyl pendant groups with AlEt₃ leads to the macroinitiator for the ROP of ε-CL, γ-substituted ε-CL, lactides, etc. Therefore, comb-like polyesters, graft copolymers and hyperbranched (dendritic) (co)polyesters can be prepared,[6,7] as illustrated by scheme 4.

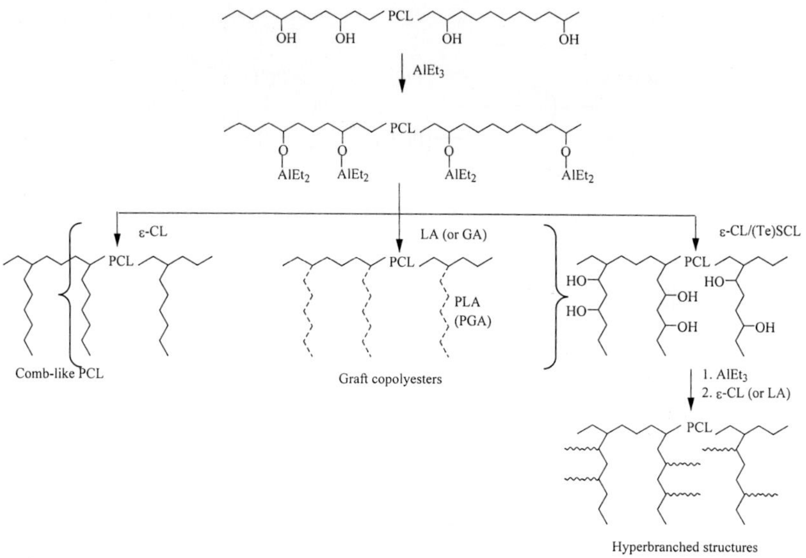

Scheme 4. Synthesis of comb-like, graft and hyperbranched (co)polyesters

In order to go a step further in the macromolecular engineering, ε-CL was copolymerized with TOSUO and (Te)SCL. An interesting characteristic of these terpolymers is the difference in stability of the protected hydroxyl groups. Indeed, the hydrolysis of the silanolate groups is selective in the presence of the acetal protecting groups. Therefore, the sequential deprotection of the hydroxyl groups followed by their conversion into Al alkoxides is an efficient strategy to prepare hetero-graft copolymers (scheme 5).

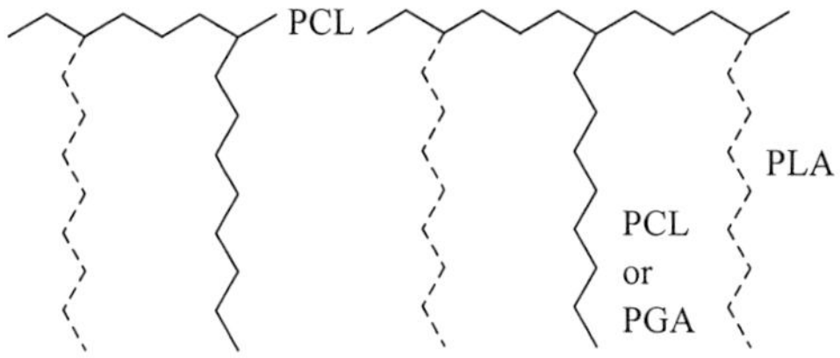

Scheme 5. Heterograft copolymers synthesized from ε-CL/TOSUO/(Te)SCL terpolymers

A variation on the theme may be found in the initiation of the ring-opening copolymerization of the ε-CL/TOSUO (or (Te)SCL) mixture by a multifunctional compound, which may lead to novel architectures in which block and graft structures, or dendrimeric and graft structures coexist (scheme 6).[8]

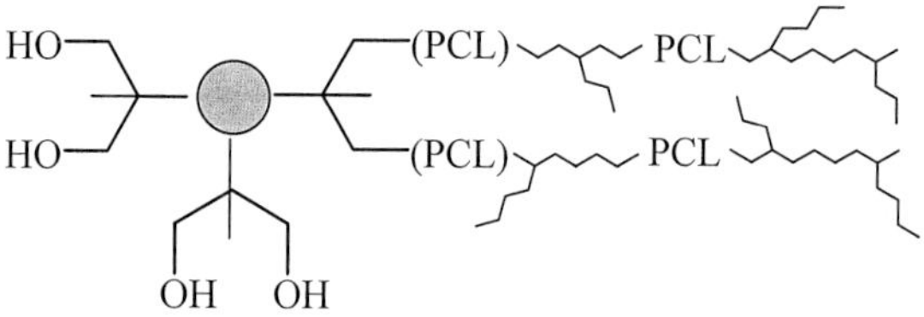

Scheme 6. Possible architectures from multifunctional ROP initiators

Additional examples of hydroxyl substituted monomers have been studied, including N-pyrrole-11-undecanol (Py-OH)[9] and 2-hydroxymethyl-5-norbornene (NBE-OH).[10]

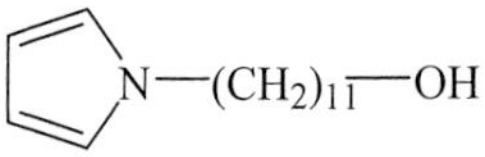

Py-OH

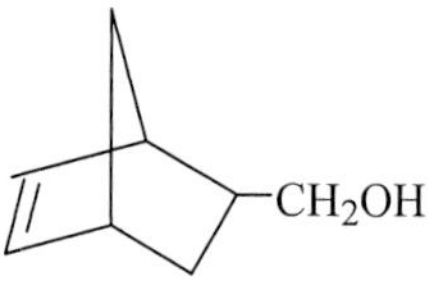

NBE-OH

These molecules are precursors of polyester macromonomers bearing a pyrrole or a norbornene polymerizable group.[9,10] The Py-PCL macromonomer was copolymerized with pyrrole (Py) by chemical or electrochemical oxidation leading to graft copolymers with improved film forming and mechanical properties compared to polyPy.[9] α-Norbornenyl PCL was also copolymerized with norbornene and norbornene acetate by the [RuCl$_2$(p-cymene)]$_2$ PCy$_3$/(trimethylsilyl)diazomethane complex (Grubbs catalyst)]. The metathesis copolymerization is better controlled in case of the norbornene acetate comonomer which is less reactive than norbornene and decreases the extent of side reactions.[10] It must be noted that the macromonomer can be homopolymerized into high molecular weight comb-like chains of low polydispersity (M_w/M_n = 1.10) within high yields (90%).[10] Finally, random copolymers of norbornene and norbornene acetate are macroinitiators for the graft copolymerization of lactones and lactides after deprotection of the hydroxyl groups (n-BuLi in THF).[11]

p-Chloromethylstyrene is a known styrene derivative which associates a polymerizable vinyl group and a benzyl chloride in the same molecule. The nitroxide-mediated controlled radical polymerization of styrene is very well-known and it proved to be tolerant for reactive derivatives including p-chloromethylstyrene and other functional monomers, such as HEMA.[12] Benzyl chloride is a potential initiator for the controlled radical polymerization by ATRP. Therefore, p-chloromethylstyrene is a dual molecule which allows two different CRP processes to be combined in one synthesis scheme. Either α-styryl macromonomers can be synthesized or macroinitiators for the "grafting-from" technique based on ATRP can be made available.

Similarly to p-chloromethylstyrene, γ-(2-bromo-2-methylpropionate)-ε-caprolactone (BMPCL) is an interesting combination of a cyclic monomer and an initiator for ATRP polymerization.

BMPCL

Chains end-capped by a polymerizable ε-caprolactone unit can thus be prepared by CRP and copolymerized by ROP with formation of graft copolymers. Moreover, copolyesters of ε-CL and BMPCL are easily synthesized and the pendant activated bromides can either initiate a "grafting-from" reaction by ATRP, or be dehydrohalogenated into acrylic-type double bonds (making the chains crosslinkable) or be quaternized (making the chains amphiphilic or water-soluble) as shown in scheme 7.[13]

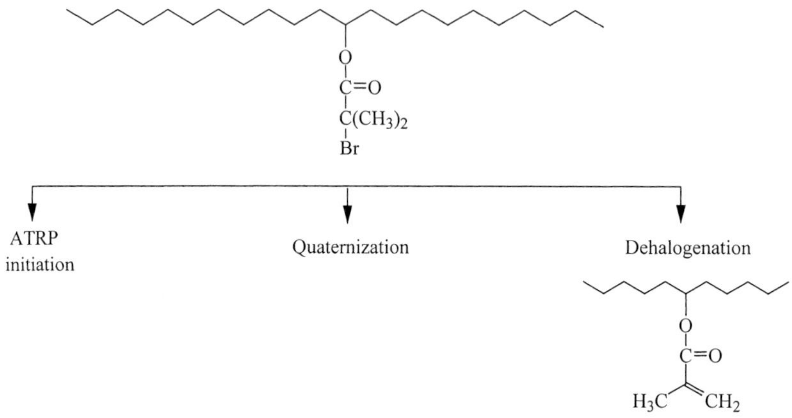

Scheme 7. Possible derivatizations of ε-CL/BMPCL copolyesters

Unsymmetrical difunctional initiators

A substantial improvement of the concept of performing dual living/controlled polymerization from a single molecule was improved by C.J. Hawker and J.L. Hedrick who proposed to use a dual initiator for the living/controlled polymerization of dissimilar monomers without the need of intermediate activation or transformation steps.[14] Coexistence of hydroxyl groups with either alkoxyamine or tribromo initiating groups was found to be compatible with the experimental conditions for both the living ROP of ε-caprolactone and nitroxide-mediated or atom transfer controlled radical polymerizations (scheme 8).

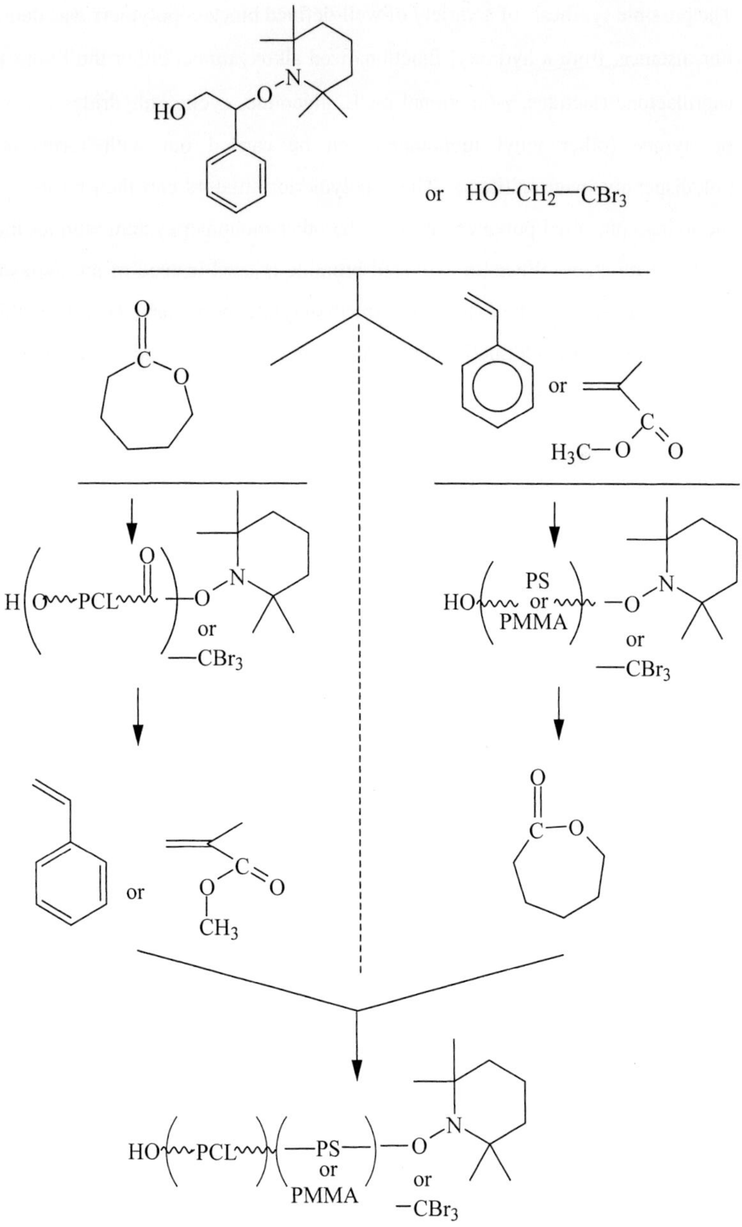

Scheme 8. Synthesis of diblocks from a dual initiatior active in CRP and ROP

The possible synthesis of a variety of well-defined block copolymers was demonstrated. For instance, from a hydroxyl-functionalized alkoxyamine, either the living ROP of ε-caprolactone (lactides, γ-functional ε-CL, glycolide, cyclic anhydrides…), or the CRP of styrene (other vinyl monomers) can be carried out with formation of low polydispersity macroinitiators. These polymeric initiators can then be used to initiate the living/controlled polymerization of the other monomer system without the need for intermediate steps. When an activated bromide is used instead of an alkoxyamine, the copolymerization can be extended to (meth)acrylate monomers. Therefore, this strategy allows novel well-defined block copolymers to be readily synthesized under nondemanding experimental conditions.

Last but not least, the sequential two-step method discussed above can be turned into a one-step process whenever the two monomers and initiating/catalytic systems are compatible and tolerate each other, and the kinetics of the two simultaneously occurring reactions is comparable at the selected temperature. This unusual one-step block copolymerization was successfully carried out in bulk, in the cases illustrated in scheme 8.[15] This one-step approach was also extended to the synthesis of graft copolymers, particularly of poly(MMA-graft-polyε-CL) (scheme 9). The simultaneous CRP of a mixture of MMA and HEMA was initiated by 2,2-dichloroacetophenone and catalyzed by RhCl(PPh$_3$)$_3$, and the ROP of ε-CL was promoted by the hydroxyl group from HEMA and Al isopropoxide as catalyst. The reaction was conducted at 50°C, in bulk conditions, and the molecular weight of the PMMA backbone (after complete hydrolysis of the polyester grafts) was the value expected from the monomer-to-initiator molar ratio.

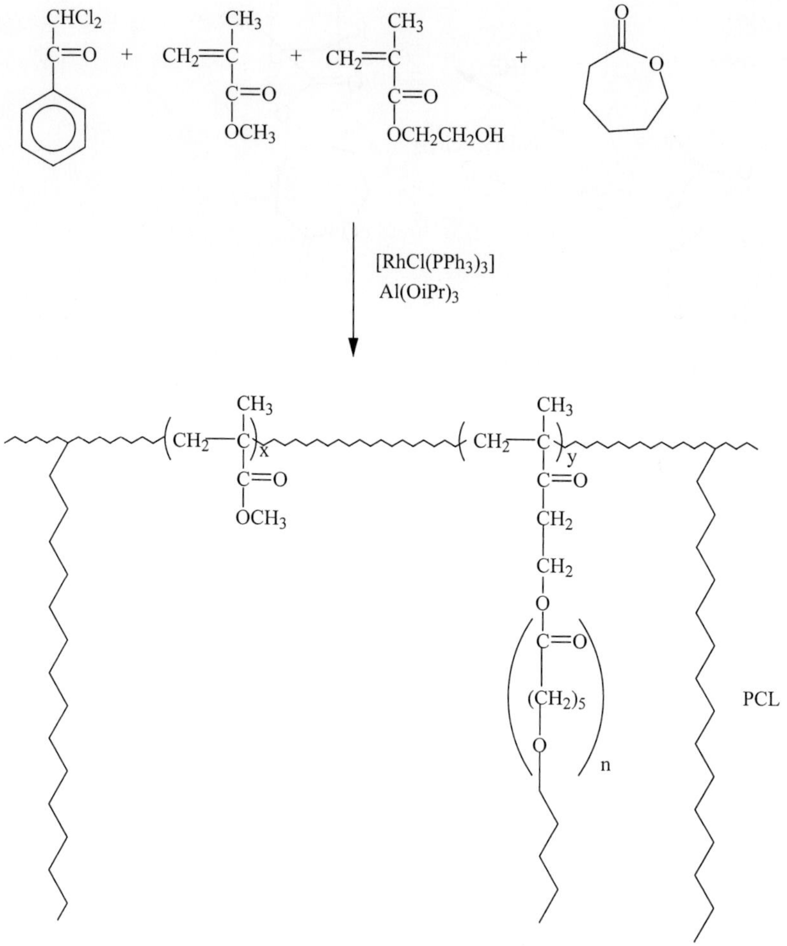

Scheme 9. One-step synthesis of poly(MMA-graft-poly ε-caprolactone)

Unsymmetrical Difunctional Monomers

4-(acryloyloxy)-ε-caprolactone (AOCL) is a novel difunctional acrylate-lactone monomer which can be selectively polymerized in a living/controlled way by two different polymerization mechanisms, i.e., ATRP and ROP (scheme 10).[16]

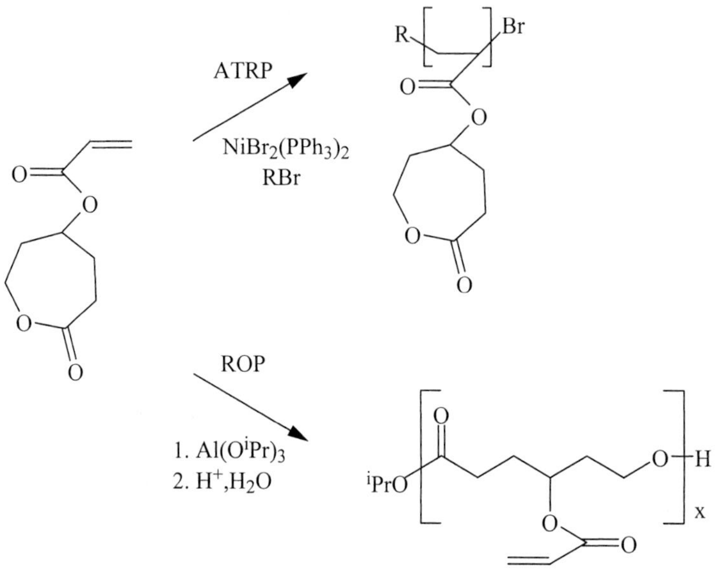

Scheme 10. Alternative polymerization routes for 4-(acryloyloxy)-ε-caprolactone

This scheme illustrates that the ATRP of AOCL leads to new polyacrylates containing pendant caprolactone groups. The molecular weight is controlled and the polydispersity is low ($M_w/M_n \sim 1.1$ in the range of 10-25000 M_n). Alternatively, ROP of AOCL makes new functional aliphatic polyesters available with a good control of the molecular parameters. The pendant acrylate groups in the copolyester chains are sensitive to thermally or photochemically generated radicals, which opens the way to cross-linkable biodegradable materials.

Norbornenylmethylene acrylate (NBE-A) is another example of unsymmetrical difunctional monomer, that has been recently electropolymerized (through the acrylate double bond) and polymerized by ring-opening metathesis (ROMP).[17] It must be pointed out that the electroreduction of monomers of the (meth)acrylate type at an appropriate potential leads to the chemisorption of a thin polymer film onto the cathode.[18,19] Moreover, well-defined and highly active catalysts in ROMP of cycloolefins were recently reported by Schrock[20] and Grubbs.[21] Therefore, the electrografting process was combined with living ROMP in order to overcome the intrinsic limitations of electrochemistry, i.e. very low thickness of the "electrografted-from" film (< 100 μm) due to the rapid passivation of the electrode and restriction to (meth)acrylates that contain no protic functions. Scheme 11 illustrates the strategy that was worked out.

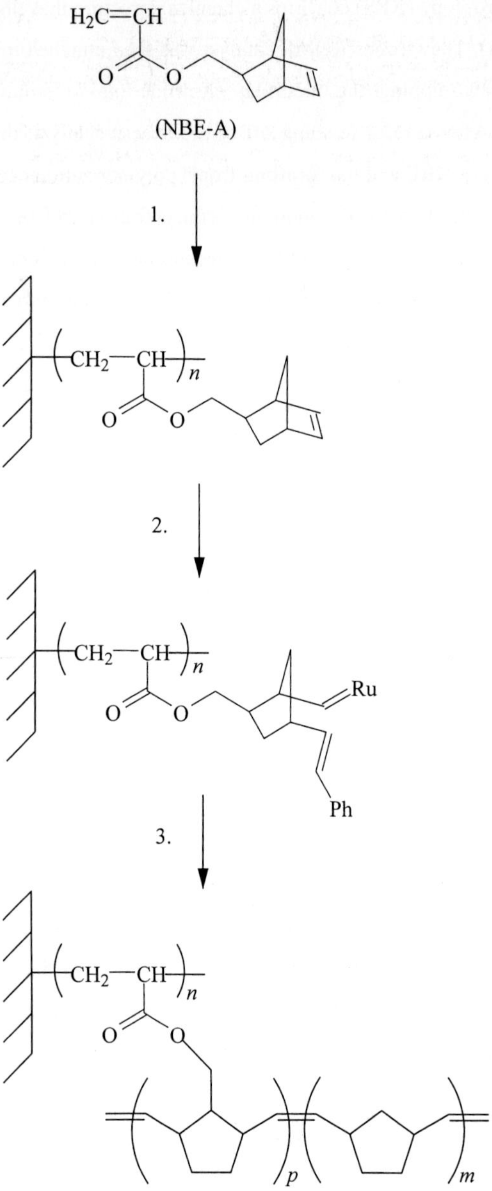

H₂C=CH
(NBE-A)
1.
2.
Ru
Ph
3.

Scheme 11. Combination of electrografting and ROMP for the dual NBE-A monomer

Upon cathodic electropolymerization of NBE-A at a well-defined potential, the cathode is passivated within a few seconds. The analysis of the insulating film by X-ray Photoelectron Spectroscopy (XPS) confirms a chemical structure that fits the expectedly grafted poly(NBE-A). The Grubbs ROMP catalyst was then attached to the poly(NBE-A) coated surface by dipping the modified electrode into a solution of this Ru containing catalyst. After careful washing off of the excess catalyst, the substrate was exposed to a solution of NBE and the "grafting from" polymerization occurred. The film thickness is easily controlled by the monomer concentration and the polymerization time. This mixed electropolymerization/ROMP technique is very versatile, e.g., it can be carried out on metallic (steel, copper…) plates and wires, and on carbon plates and fibers. The brush (of polyNBE, in scheme 11) which is grafted from the primary polyacrylate coating can be changed depending on the pendant "monomer" attached to this polyacrylate coating and the chemistry worked out for the implementation of the "grafting from" step. The adhering top coating can thus be designed for imparting specific functionality to the surface, e.g. for sensor development.

The unsymmetrical 3-(2-acryloyloxyethyl)thiophene and N-(2-acryloylethyl(pyrrole)) monomers were recently synthesized with the purpose to prepare electrically conducting polymers strongly adhering to metallic surfaces by the aforementioned electrochemical technique.[22] Scheme 12 shows the two-step procedure which consists of the cathodic electrografting of polyacrylate chains bearing a precursor of the conducting polymer in the ester group (thiophene, in scheme 12), followed by the anodic polymerization of this precursor (in the presence or not of additional monomer (i.e. thiophene) in solution). The strongly adhering two-component film is conducting and electroactive, as shown by cyclic voltammetry (reversible doping and dedoping).

The unusual electrografting of acrylate containing dual monomers is thus a very promising technique to provide an organic coating with strong adhesion to inorganic (conducting) substrates with the additional flexibility in the choice of the chemical properties via the second polymerizable moiety of the monomer.

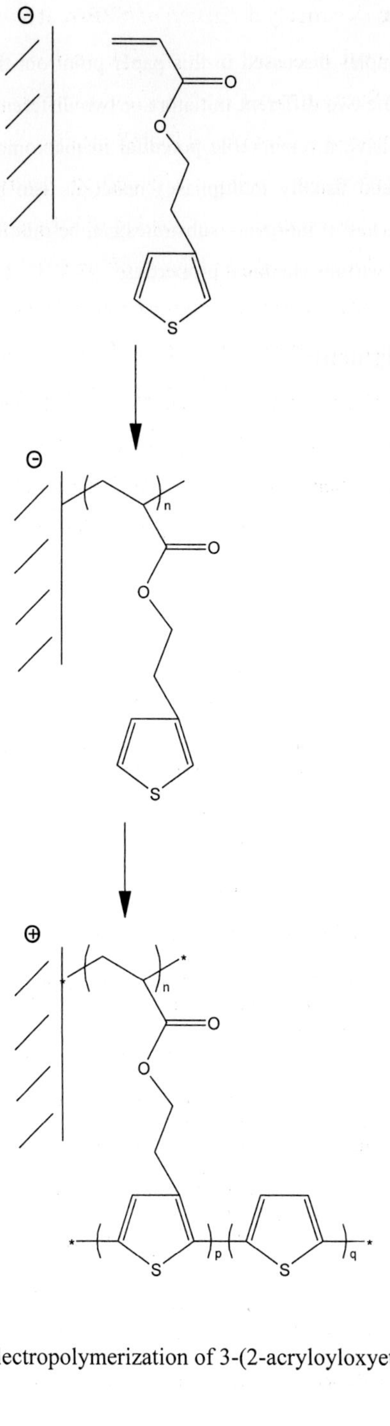

Scheme 12. Two-step electropolymerization of 3-(2-acryloyloxyethylthiophene)

58

Conclusion

The few examples discussed in this paper point out that single difunctional molecules which associate two different initiators or two different monomers or one monomer and one initiator have a remarkable potential in macromolecular engineering. Novel two-component (and usually multiphase) materials can accordingly be tailored, and the surface properties of inorganic substrates can be modified by strongly adhering organic coatings with various chemical properties.

Acknowledgment

R.J. is very much indebted to the "Services Fédéraux des Affaires Scientifiques, Techniques et Culturelles" for support to CERM in the frame of the "PAI 4-11 : Supramolecular Chemistry and Supramolecular Catalysis". He is very grateful to all the collaborators, who co-authorized the papers cited in the "References" section.

[1] (a) A. Noshay, J. E. McGrath, *"Block Copolymers: Overview and Critical Survey"*, Academic Press, New York, 1977; (b) D. C. Allport, W.H. Janes, *"Block Copolymers"*, Wiley, New York, 1973; (c) R. Jérôme, R. Fayt, T. Ouhadi, *"New Developments in Block Copolymerization"*, Progr. Polym. Sci. **1984**, *10*, 87.

[2] D. Mecerreyes, Ph. Dubois, R. Jérôme, J. L. Hedrick, *Macromol. Chem. Phys.* **1999**, *200*, 156.

[3] D. Tian, Ph. Dubois, R. Jérôme, *Polymer* **1996**, *37*, 3983; *J. Polym. Sci., Polym. Chem.* **1997**, *35*, 2295.

[4] D. Tian, Ph. Dubois, Ch. Grandfils, R. Jérôme, *Macromolecules* **1997**, *30*, 406.

[5] G. Pitt, Z. W. Gu, P. Ingram, R. W. Hendren, *J. Polym. Sci., Polym. Chem.* **1987**, *25*, 955.

[6] D. Tian, Ph. Dubois, R. Jérôme, *Macromol. Symp.* **1998**, *130*, 217.

[7] F. Stassin, O. Halleux, Ph. Dubois, Ch. Detrembleur, Ph. Lecomte, R. Jérôme, *Macromol. Symp.* **2000**, *153*, 27.

[8] M. Trollsas, J. L. Hedrick, D. Mecerreyes, Ph. Dubois, R. Jérôme, H. Ihre, A. Hult, *Macromolecules* **1998**, *31*, 2756.

[9] C. Jérôme, L. Martinot, P. Louette, R. Jérôme, *Macromol. Symp.* **2000**, *153*, 305.

[10] D. Mecerreyes, D. Dahan, Ph. Lecomte, Ph. Dubois, A. Demonceau, A. F. Noels, R. Jérôme, *J. Polym. Sci., Polym. Chem.* **1999**, *37*, 2447.

[11] Ph. Lecomte, D. Mecerreyes, Ph. Dubois, A. Demonceau, A. F. Noels, R. Jérôme, *Polymer Bull.* **1998**, *40*, 631.

[12] C.J. Hawker, D. Mecerreyes, E. Elce, J. Dao, J.L. Hedrick, I. Barakat, Ph. Dubois, R. Jérôme, W. Volksen, *Macromol. Chem. Phys.* **1997**, *198*, 155.

[13] C. Detrembleur, M. Mazza, X. Lou, O. Halleux, Ph. Lecomte, D. Mecerreyes, J. L. Hedrick, R. Jérôme, *Macromolecules* **2000**, *33*, 7751.

[14] C. J. Hawker, J. L. Hedrick, E. E. Malmström, M. Trollsas, D. Mecerreyes, G. Moineau, Ph. Dubois, R. Jérôme, *Macromolecules* **1998**, *31*, 213.

[15] D. Mecerreyes, G. Moineau, Ph. Dubois, R. Jérôme, J. L. Hedrick, C. J. Hawker, E. E. Malmström, M. Trollsas, *Angew. Chem. Int. Ed.* **1998**, *37*, 1274.

[16] D. Mecerreyes, J. Humes, R. D. Miller, J. L. Hedrick, C. Detrembleur, Ph. Lecomte, R. Jérôme, J. San Roman, *Macromol. Rapid Commun.* **2000**, *21*, 779.

[17] C. Detrembleur, C. Jérôme, M. Claes, P. Louette, R. Jérôme, *Angew. Chem.*, accepted for publication.

[18] J. Tanguy, G. Deniau, C. Augé, G. Zalezer, G. Lécayon, *J. Electroanal. Chem.* **1994**, *377*, 115.

[19] N. Baute, Ph. Teyssié, L. Martinot, M. Mertens, Ph. Dubois, R. Jérôme, *Eur. J. Inorg. Chem.* **1998**, 1711.
[20] G. Bazan, J. Oskam, H. Cho, L. Park, R. Schrock, *J. Am. Chem. Soc.* **1991**, *113*, 6899.
[21] P. Schwab, M. France, J. Ziller, R. Grubbs, *Angew. Chem., Int. Ed. Engl.* **1995**, *34*, 2039.
[22] D. E. Labaye, C. Jérôme, V. M. Geskin, P. Louette, R. Lazzaroni, L. Martinot, R. Jérôme, to be submitted.

Defined Synthesis of Copolymers Using Metallocene Catalysis

Walter Kaminsky and Georg Schupfner*

Department of Chemistry, University of Hamburg, Bundesstr. 45,
D-20146 Hamburg, Germany

Summary: The copolymerization of propene with small amounts of ethene, catalyzed by tetrahydroindenyl zirconocenes such as [En(H$_4$Ind)$_2$]ZrCl$_2$ or [Me$_2$Si(H$_4$Ind)$_2$]ZrCl$_2$ and MAO in liquid propene produces polymers with much higher activities and molecular weights than the homopolymerization of propene. The normal bisindenyl complexes doesn't present such differences. The investigation of the microstructure shows for the tetrahydroindenyl catalyst that after a 2,1-insertion of a propene unit the system is in a sleeping state and can be activated when an ethene unit is inserted. In this case these catalysts become faster than the *ansa* bis-indenyl catalysts. An active catalyst for the copolymerization of ethene and norbornene is the more temperature stable [Me$_3$PhPen(Flu)]ZrCl$_2$. This catalyst produces atactic copolymers with high molecular weights of over 900 000 g/mol at 30 °C and 38 mol % of norbornene content.

Introduction

Metallocenes are highly active catalysts for the production of precisely designed polyolefins and copolymers. Especially zirconocene methylalumoxane (MAO) catalysts, half-sandwich amido titanium complexes in a combination with perfluorophenylborate have opened a frontier in the area of new copolymer synthesis and processing. The copolymers obtained show different microstructures, tacticities and properties[1-3].

Important are ethylene-1-octene and ethylene-styrene copolymers. These polymers show increased impact strength and toughness, better melt characteristics or elasticity, and improved clarity in films[4]. Supporting of the zirconocenes on silica decreases the necessary surplus of MAO and can change the tacticity[5].

Metallocenes are useful catalysts for the production of cycloolefin copolymers (COC) and α-olefin copolymers – new types of polymers with special properties and a high potential as engineering plastics[6-8]. Ethene/norbornene copolymers are the most interesting for technical uses because of the easily available monomers. Due to different incorporation values of the cyclic olefin in the copolymer, the glass transition temperature can vary over a wide range independenly from the used catalysts. A copolymer with 50 mol % of norbornene yields a material with a glass transition point of 145 °C. A Tg of 205 °C can be achieved at higher incorporation rates. The metallocene [Me$_2$C(tert-BuCp)(Flu)]ZrCl$_2$ shows not only high activities for the copolymerization of ethene with propene or norbornene, and gives alternating structure, too[9]. The melting point of the alternating copolymer depends on the molar ratio of norbornene in the polymer while the glass transition temperature is almost independent. A maximum melting point of 320 °C was reached. Crystallinity of these copolymers is detected using wide angle X-ray scattering of films. The degree of crystallinity is about 22 % for copolymers containing 49 mol% of the cycloolefin. Block structures can be prepared also[10].

Ethene can be copolymerized with propene, dienes and other olefins to give EP or EPDM elastomers. It was shown in the past that with some ansa zirconocenes the activities of the copolymerization of ethene with propene increase. Investigations by Busico et al. have shown that sleeping active centers, formed by 2,1-insertion of propene, could be activated by hydrogen or ethylene[11-13].

Experimental Part

Different C$_1$- and C$_2$-symmetric zirconocenes were used for the copolymerization of the olefins (Fig. 1)

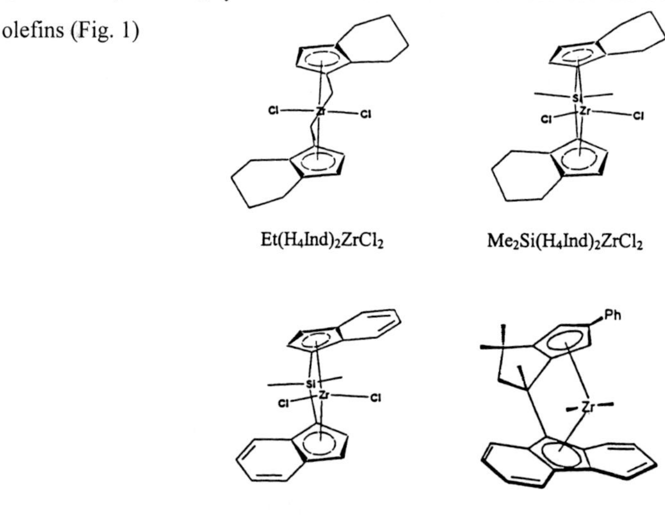

Fig. 1. Molecular structures of the used zirconocenes and their terms

The zirconocenes were prepared as described in the literature[14,15]. Polymerizations were carried out under an argon atmosphere using a 1 l Büchi A6 Type I autoclave equipped with an additional external cooling system. For COC experiments, the reactor was evacuated at 95 °C for 1 h and charged subsequently with 200 ml toluene, 500 mg MAO(purchased by Witco), norbornene, and ethene at different pressures. Norbornene was dried over triiso-butylaluminum and subsequently distilled. Gaseous monomers (ethene, propene) were purified by passing through columns with a copper-catalyst (BASF R3-11) and molecular sieve of 3-4 Å. Copolymerizations of propene with ethene were carried out in 375 ml of liquid propene. The polymerization was started by injection of a toluenic solution of the metallocene. The polymerization was quenched by addition of 5 ml of ethanol. Work-up proceeded by stirring over night in diluted hydrochloric acid followed by neutralization with aqueous $NaHCO_3$ and washing with water. After phase separation, the polymer was precipitated, if possible. Otherwise the organic solvent was removed under reduced pressure and the obtained polymer was dried in vacuo.

All ^{13}C NMR spectra of the polymers were recorded on a Bruker MSL 300 spectrometer operating at 75.47 MHz and 100 °C. Polymer samples were dissolved in perchlorobutadiene and tetrachloroethane-d_2. Molar masses and molar mass distributions were determined by size exclusion chromatography on a Waters 150-C instrument (1,2,4-trichlorobenzene at 135 °C) employing a PL-EMD-960 evaporation light scattering detector. Additional molar mass determination was conducted by viscosimetry using an Ubbelohde viscosimeter at 30 °C (Kapillare Oa, K = 0,005).

Results and Discussion

Copolymerization of Propene with Ethene

We were the first to show that small amounts of ethene added to propene increase the activity of the polymerization when using tetrahydroindenyl zirconocenes (see Fig. 1)[16]. With these catalysts up to 2 mol% in a 2,1-insertion of propene is observed (Fig. 2).

After a rare 2,1-insertion of propene into the active site, the zirconium-$CH(CH_3)$-bond is sterically so hindered that a next propene insertion is much slower [sleeping state (a)]. But it is easy to incorporate an ethene unit into this bond (b). After this step it is possible again to insert a propene unit in the normal 1,2-position (c) high rate.

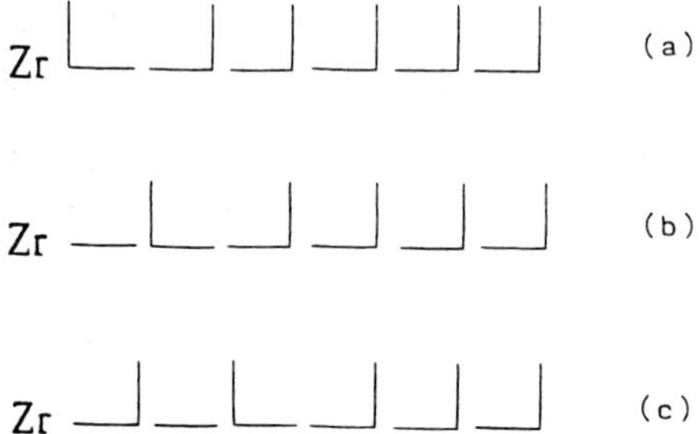

Fig. 2. 2,1 Insertion of propene in a Zr-polymer bond (a) followed by an ethene insertion (b) and a 1,2-propene insertion (c)

The activity of the copolymerization of propene with small amounts of ethene can be up to fivetimes higher than the homopolymerization (Tab. 1).

Table 1. Propene/ethene copolymerization with different zirconocene/MAO catalysts in liquid propene at different temperatures; MAO=500 mg, 375 ml liquid propene, ethene flow =1,7 l/h, a): the activity of homopolypropene, produced under same conditions for comparison.

Catalyst	Polymerizations-temperature (°C)	Zircono-cene ($\times 10^6$ mol)	Activity copolymer kg_{Pol}/mol_{Zr}	Activity a) homopolymer kg_{Pol}/mol_{Zr}
$[Me_2Si(H_4Ind)_2]ZrCl_2$	0	8,05	650	21
	30	2,00	16 200	
	30	0,43	20 330	4 290
	60	0,09	149 200	71 700
$[Et(H_4Ind)_2]ZrCl_2$	0	3,00	1 200	750
$[Me_2Si(Ind)_2]ZrCl_2$	0	6,13	520	320
	30	0,62	9 900	8 000
	60	0,07	55 960	51 200

Astonishingly, the increase of the activity is only given for tetrahydroindenyl complexes. The bisindenyl metallocene $[Me_2Si(Ind)_2]ZrCl_2$ shows nearly no 2,1-insertion and by addition of small amounts of ethene only a very small increase of the activity. To explain the increase of the activity by a factor of two it could be calculated that for the homopolymerization of propene about half of all active centers are "sleeping".

The 2,1-insertion of propene blocks not only the following insertion but favours β-hydrogen transfer reactions and therefore chain terminations. By this means the molecular weight of the obtained homo polypropene is low. The copolymer obtained with 2-6 mol% of ethene units has a significant higher molecular weight. Because of the higher activity for the copolymerization and the same rate for the chain termination, the chain length of the polymer is longer (Tab. 2).

Table 2. Molecular weights of propene/ethene copolymers with 2-6 mol% of ethene prepared by different catalysts. Polymerization conditions are the same as in Tab. 1. M_w homopolymer: molecular weight of polypropene without ethene addition.

Catalyst	Temp. (°C)	M_w (kg/mol)	M_n (kg/mol)	M_w/M_n	M_w homo polymer
$[Me_2Si(H_4Ind)_2]ZrCl_2$	0	239	79	3,0	155
	30	71	31	2,3	45
	30	109	43	2,5	
	60	26	12	2,2	18
$[Et(H_4Ind)_2]ZrCl_2$	0	257	65	2,7	
$[Me_2Si(Ind)_2ZrCl_2$	0	113	45	2,6	122
	30	88	37	2,2	75
	60	48	20	2,4	48

The catalyst $[Me_2Si(Ind)_2]ZrCl_2$/MAO gives nearly the same molecular weights for the homo- and copolymerization. Fig. 3 shows that the GPC curves become more narrow with increasing polymerization temperatures.

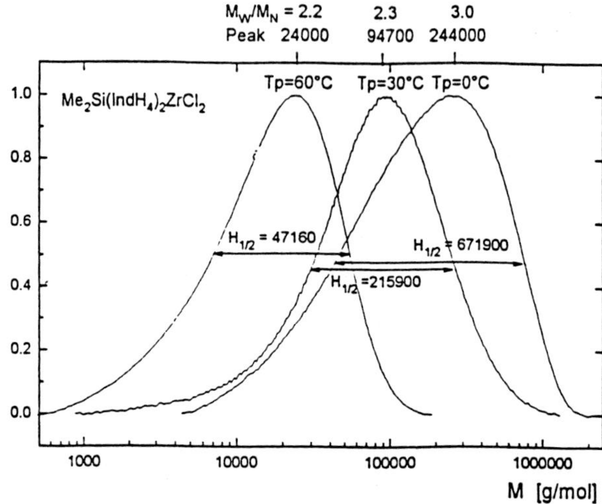

Fig. 3. GPC curves of propene/ethene copolymers with 2 mol% of ethene catalyzed by [Me$_2$Si(H$_4$Ind)$_2$]ZrCl$_2$/MAO

The microstructures of the propen/ethene copolymers were determined by ^{13}C NMR measurements using the calculation of Cheng and Benett[17]. Table 3 presents the calculated and measured chemical shifts for the 2,1-insertion followed by an ethene insertion in a propene/ethene copolymer.

Table 3. ^{13}C-NMR chemical shifts (ppm) of propene/ethene copolymers for the ethene insertion after 1,2- and followed by 1,2-insertion of propene (e-i), for the 2,1-insertion of propone followed by an ethene insertion (2,1–e-i-), for the meso 2,1-insertion followed by a 1,2-insertion of propene (2,1-i); (P) = primary carbon atom, (S) = secondary carbon atom, (T) = tertiary carbon atom.

Sequence (e-i)	1(S)	2(T)	3(P)	4(S)	5(S)	6(S)	7(T)
calculated	45,9	31,0	20,9	37,7	24,5	37,7	31,0
experimental	45,9	30,7	20,8	37,6	24,3	37,6	30,7
Sequence (2,1-e-i)	1(S)	2(T)	3(P)	4(S)	5(S)	6(T)	7(P)
calculated	45,7	31,3	21	34,8	34,5	34	20,3
experimental	45,5	31,0	21	34,6	34,3	33,8	20,1
Sequence (2.1-i)	1(S)	2(T)	3(P)	4(T)	5(P)	6(S)	7(T)
calculated	41,9	35,9	17,8	38,4	17,1	30,2	31,4
experimental	41,9	35,6	17,4	38,3	17,8	30.1	31,2

The NMR-measurements show that after a 2,1-insertion of propene selectively ethene is inserted. 2,1-inserted propene units surrounded by 1,2-inserted units are no longer detectible (Tab. 4).

Unexpected is the fact that for copolymers with [Me$_2$Si(H$_4$Ind)$_2$]ZrCl$_2$/MAO at high temperatures the amount of 2,1- and 1,3-insertions of propene are reduced significantly. This shows that in the presence of ethene as a comonomer misinsertions are reduced.

Table 4. ^{13}C-NMR spectroscopic determination of the microstructures of propene/ethene copolymers compared to homo polypropene. Explanations for the symbols see Tab. 3. 1,3-i: 1,3-insertion of propene, I: isotacticity mmmm pentads, incorporations in mol%.

Catalyst	Temp. (°C)	Copolymers I mmmm	e-i	2,1-e-i	2,1-i	Ethen	Homopolypropene I	2,1-i	1,3-i mmmm
[Me$_2$Si(H$_4$Ind)$_2$]-ZrCl$_2$	0	95	1,6	0,5	0	2,1	94	0,6	0
	30	91	2,3	0,3	0	2,4	96	0,3	0,2
	60	85	3,0	0	0	0	3,0	91	0
0,5									
[Me$_2$Si(Ind)$_2$]ZrCl$_2$	0	92	2,7	0	0	2,6	96	0,3	0
	30	93	5,1	0,3	0	5,4	93	0,4	0
	60	83	5,8	0,8	0	5,6	91	0,5	0

Ethene-Norbornene Copolymers

To increase the thermostability and the molecular weight of cycloolefin copolymers, a new metallocene with a more stable bridge was used[15]. For the propene polymerization we have received good results with the pentalene-fluorenyl zirconium complex (Me$_3$PhPen(Flu)ZrCl$_2$) showed in Fig. 1. Ethene and norbornene were polymerized with this complex in combination with MAO under different conditions. Table 5 shows composition and properties of the copolymers produced at variable norbornene concentrations.

Table 5. Copolymerization of ethene with norbornene by [Me$_3$PhPen(Flu)ZrCl$_2$]/MAO at 30 °C. x_N : norbornene mol part in the starting solution; X_N: norbornene mol part in the copolymer; c_E: ethene concentration: 0,237 mol/l; Zr-concentration: 5 · 10^{-6} mol/l; MAO: 2,5 g/l; Tg: glass transition temperature, Tm: melting point; M_w: molecular weight; n.d.: not detected.

x_N	X_N	Tg (°C)	Tm (°C)	M_w (g/mol)	Activity (kg $_{Copo}$/mol $_{Zr}$·h·c$_E$)	
0	0		135	660 000	3	700
0,20	0,046		98	658 000	5	780
0,37	0,094		66	538 000	7	600
0,54	0,159	2	atactic	593 000	6	700
0,79	0,280	65	atactic	758 000	2	500
0,91	0,380	93	atactic	930 000	n.d.	

Copolymers with a norbornene content of over 15 mol% are amorphous as also measured for metallocene catalysts[18].

Copolymers with lower norbornene contents are partial crystalline with melting points between 66 and 135 °C. Due to different incorporation rates of the cyclic olefin into the copolymer, the glass transition temperature varies over a wide range. A copolymer containing 38 mol% of norbornene yields a material with a glass transition point of 93 °C.

The molecular weights are unusually high and increase with the norbornene concentration in the starting solution. Normally the molecular weight decreases with a higher norbornene concentration. Even at an incorporation of 38 mol% of norbornene the copolymer has a molecular weight of 930 000 g/mol. A minimum molecular weight is given at about 10 mol% of norbornene. The activity reaches 7 600 kg $_{copo}$/mol $_{Zr}$·h by 30 °C and can be increased at higher temperatures.

The high molecular weight is very useful for the physical properties and industrial applications. These COC materials present excellent transparency and very high service temperatures. They are soluble, chemically resistant and can be melt-processed. Their stability against hydrolysis and chemical degradation in combination with their stiffness let them become desirable materials for optical applications, e.g. for compact discs, lenses, optical fibers and films. The first commercial COC plant runs at TICONA with a capacity of 30 000 tons a year.

This shows that it is possible to find metallocene/MAO catalysts which can produce copolymers with high molecular weights even by high temperatures and norbornene concentrations.

References

[1] J. Scheirs, W. Kaminsky (eds.) Metallocene-Based Polyolefins Vol. I + II, Wiley (2000), Chichester
[2] R. Blom, A. Follestad, E. Rytter, M. Tilsel, M. Ystenes (eds.) Organometallic Catalyst and Olefin Polymerization, Springer 2001, Berlin
[3] S.D. Ittel, L.K. Johnson, M. Brookhart, Chem. Rev. 2000, 100, 1169
[4] A. Torres, K. Swogger, C. Kao, S. Chum, in Reference 1, p. 143
[5] W. Kaminsky, F. Renner, Makromol. Chem. Rapid Commun. 1993, 13, 239
[6] W. Kaminsky, A. Bark. M. Arndt, Makromol. Chem., Macromol. Sym. 1991, 47, 8
[7] H. Cherdron, M.-J. Brekner, F. Osan, Angew. Makromol. Chem. 1994, 223, 121
[8] D. Ruchatz, G. Fink, Macromolecules 1998, 31, 4669
[9] W. Kaminsky, M. Arndt, I. Beulich, Polym. Mater. Sci. Eng. 1997, 76, 18
[10] M. Arndt, I. Beulich, Macromol. Chem. Phys. 1998, 199,1221
[11] V. Busico, R. Cipullo, P. Corradini, Makromol. Chem. Rapid Commun. 1993, 14, 97
[12] S. Liu, R. Kravchenko, R. Waymouth, J. Molecular Cat. A. Chemical 2000, 158, 423
[13] V. Busico, R. Cipullo, Progress in Polymer Science 2001, 26, 443
[14] F.R.W.P. Wild, L. Zsolnai, G. Huttner, H.H. Brintzinger, J. Organomet. Chem. 1982 232, 233
[15] R. Werner, Dissertation Hamburg 1999
[16] G.U. Schupfner, Dissertation Hamburg 1995
[17] H.N. Cheng, M.A. Benett, Makromol. Chem. 1985, 188, 135
[18] W. Kaminsky, J. Chem. Soc. Dalton Trans. 1998, p. 1413

High Molecular Weight Polypropene Elastomers via "Dual-Side" Zirconocene Dichlorides

*Jürgen Kukral and Bernhard Rieger**

Department for Materials and Catalysis, University of Ulm, D-89069 Ulm, Germany

Summary: A series of asymmetric *ansa*-zirconocene dichlorides were investigated in propene polymerization reactions with respect to monomer concentration and polymerization temperature. The two different coordination sides of these "dual-side" catalysts allow a precise control of the total amount of randomly distributed isolated stereoerrors along the isotactic polymer chains leading to bulk properties which can be adjusted from flexible, semicrystalline to excellent thermoplastic elastic. The obtained homopolypropenes have a great potential to come into competition with commercialized copolymers for possible industrial applications.

Introduction

The development of new C_2-symmetric metallocenes for the isospecific polymerization of propene increased rapidly since Brintzinger's first introduction of homogeneous *ansa*-metallocene catalysts.[1,2] To a lesser degree, research was focused on asymmetric metallocene species. However, those structures proved to be excellent tools for the design of different polymer properties by introdcution of substitutents in key positions of the ligand framework. Ewen et al.[3] demonstrated with the synthesis of a series of different C_S-symmetric Zr(IV)-complex based on the [2-(9-η^5-fluorenyl)-2-(η^5-cyclopentadienyl)propane]zirconium dichloride that the two different coordination sides available for asymmetric metallocenes customize a broad range of material properties in migratory polyinsertion reactions of propene by modification of the polymer microstructure. So, the stereochemistry of the polymerization reaction can be changed from the production of syndiotactic polypropene to a hemiisotactic form and even to a highly isotactic material by opening or blocking selective and nonselective sides in successive insertion steps. Chien et al.[4] observed at about the same time, that elastic polypropene of narrow molecular weight distribution can be accessed by the asymmetric mono-carbon-bridged titanocene [MeHC(Me$_4$Cp)(Ind)]TiCl$_2$ resulting in increased exertions in research on suitable catalysts for the production of elastomeric polypropene.[5,6] The elastic properties were ascribed to arise from blocklike structures composed of isotactic and atactic sequences.[4,7] Our initial forays in the field of asymmetric metallocenes were pointed on the synthesis of the ethylene bridged diastereomeric *rac*-[1-(9-η^5-fluorenyl)-2-(η^5-indenyl)phenylethane]zirconium dichlorides, which contain a bulky phenyl group as bridge substituent to stabilize the conformation of the metallacycles. Therewith we realized that the δ-forward and λ-backward conformers behave completely different with respect to the stereoselectivity of the C-C-bond forming process.[8,9]

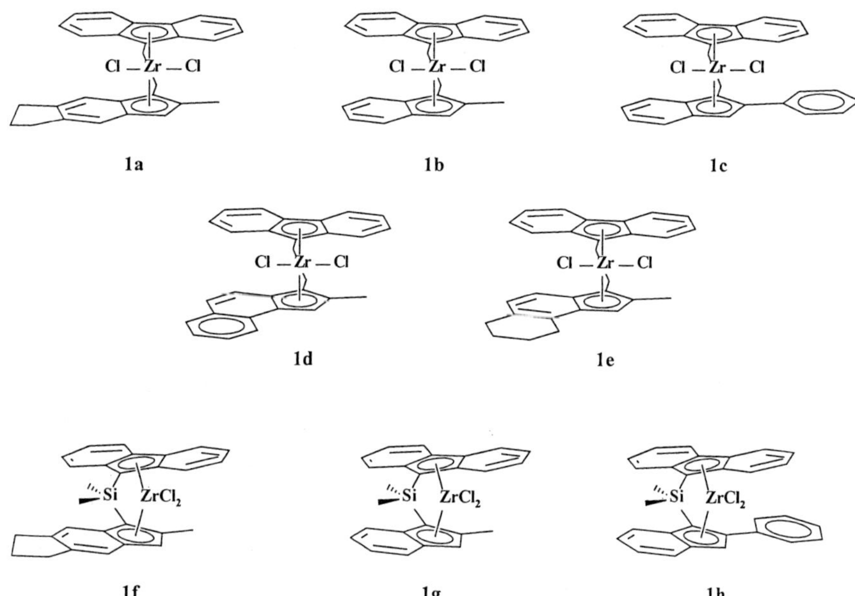

Figure 1. Summary of complexes used for the propene polymerization reactions.

In our present work we compare the polymerization behavior of a series of ethylene and dimethylsilane bridged asymmetric zirconocene dichlorides **1a-h** (Figure 1) and present information regarding the influence of the ligand substitution pattern on the polypropene characteristics with the intention to enable a tayloring of homopolypropene plastomers and elastomers in future approaches. This directive in metallocene chemistry is attributed to the new family of "dual-side" complexes, combining an isoselective side with a less selective side leading predominantly to single stereoerrors within one well defined species.[10] Asymmetric metallocenes serve as a new tool to control the amount of stereoerrors along an isotactic polymer chain allowing to adjust polymers of variable crystallinity with material properties from flexible, semicrystalline to excellent thermoplastic elastic.

Results and Discussion

Propene Polymerizations

Activity. Monomer concentration and temperature variation have proved to be efficient tool to take influence on the polymer microstructure and to achieve defined material properties .[10] Large differences were observed in the polymerization behavior of the examined catalysts by variation of these parameters and the catalyst architecture (Table 1). The highest polymer yields of all asymmetric complexes of the present study could be obtained with the ethylene bridged metallocenes **1a**/MAO (at T_P = 30 °C up to 2.9 × 10^3, at T_P = 50 °C up to 12.9 × 10^3 and at T_P = 70 °C up to 31.4 kg of PP (mol of Zr [C_3] h)$^{-1}$). The angular 4,5-substituted metallocenes **1d,e** show lower polymerization activities, which probably result from the enhanced steric encumbrance hindering the incoming propene molecule to coordinate on the active Zr-center. The activity of the complex **1b**/MAO, bearing only the 2-methyl substitution in the ligand framework, is about the same order of magnitude as for the angular complexes **1d,e**/MAO. Interestingly, the propene consumption of the ethylene bridged complexes increases within the first 15–30 minutes of the polymerization experiment (Figure 2, **1a,b**/MAO), indicating a slow activation reaction with MAO, presumably resulting from the sterically demanding substituents around the zirconium center. We found further, that the rate of catalyst activation for this type of asymmetric zirconocenes rises with increasing polymerization temperature (Figure 2, **1a,b**/MAO).[11]

Table 1. Polymerization results of propene with asymmetric complexes **1a-h**/MAO

run	cat.	amount[a]	T_p[b]	$[C_3]$[c]	t_p[d]	yield[e]	activity[f]	M_w[g]	M_w/M_n	[mmmm][h]
1	1a	15	30	1.1	42	31.9	2900	50	1.9	47.8
2	1a	6	30	1.2	73	17.3	2000	55	1.8	42.5
3	1a	3.8	30	3.0	39	18.0	2400	89	1.8	30.3
4	1a	7.5	30	3.9	33	36.9	2300	120	2.0	31.3
5	1a	2.7	30	5.1	30	17.6	2600	125	1.9	27.8
7	1a	5.0	50	0.4	46	13.9	9000	17	1.9	72.1
8	1a	3.8	50	1.1	72	51.4	10200	30	2.2	63.9
9	1a	2.5	50	1.9	50	51.2	12900	34	2.1	59.6
10	1a	2.5	70	1.1	42	60.5	31400	17	2.0	64.0
11	1b	5	30	1.2	72	15.6	2100	62	1.8	17.6
12	1b	5	30	3.0	39	16.7	1700	81	1.9	13.4
13	1b	5	30	5.1	25	16.6	1600	83	1.9	9.7
14	1c	10	30	3.0	27	18,4	1400	62	2.3	98.0
15	1c	10	30	5.1	29	20.3	800	86	2.5	86.0
16	1d	2.5	30	1.1	103	6.1	1300	47	2.9	71.9
17	1d	6.5	30	2.9	43	44.2	3300	60	2.4	69.1
18	1d	2.5	50	0.4	52	4.8	5500	13	1.8	72.6
19	1d	2.5	50	1.1	31	9	6300	27	2.0	74.4
20	1d	2.5	50	1.9	65	40	7800	38	2.0	74.7
21	1d	2.5	70	1.1	53	29.9	12300	23	3.0	66.3
22	1e	2.0	30	1.1	62	2.5	100	33	2.0	69.8
23	1e	12.8	30	2.9	46	58.8	2100	39	1.9	53.7
24	1e	2.0	50	0.4	53	1.9	2100	11	1.8	79.5
25	1e	2.0	50	1.1	29	6.7	5000	15	2.0	78.0
26	1e	2.0	50	1.9	39	17.7	5700	30	1.8	78.4
27	1e	2.0	70	1.1	45	14.4	7000	14	2.0	78.7
28	1f	10	30	1.2	120	12.5	n. d.[i]	103	1.8	61.6
29	1f	10	30	3.0	61	15.2	n. d.	132	1.9	50.6
30	1f	10	30	5.1	30	17.4	n. d.	158	1.9	44.0
31	1f	10	50	1.9	31	11.6	n. d.	64	1.8	68.0
32	1f	10	50	3.0	25	13.2	n. d.	85	1.9	67.4
33	1g	20	30	1.2	106	16.1	n. d.	73	1.8	44.1
34	1g	10	30	3.0	69	15.7	n. d.	81	2.0	39.3
35	1g	10	30	5.1	33	16.4	n. d.	92	2.0	35.3
36	1g	10	50	3.0	65	12.6	n. d.	64	1.9	56.2
37	1h	50	30	1.2	193	8.5	n. d.	32	1.9	70.5
38	1h	30	30	3.0	118	8.9	n. d.	74	1.8	62.2
39	1h	30	30	5.1	125	16.2	n. d.	99	1.8	57.9
40	1h	30	50	3.0	101	5.2	n. d.	31	1.9	71.7

[a] [μmol] (Al/Zr = 2000). [b] [°C]. [c] [mol L^{-1}]. [d] [min]. [e] [g]. [f] [kg of PP (mol of Zr [C$_3$] h)$^{-1}$. [g] [kg mol^{-1}]. [h] [%]. [i] n. d. = not determined, because of rapid deactivation of the catalyst active species.

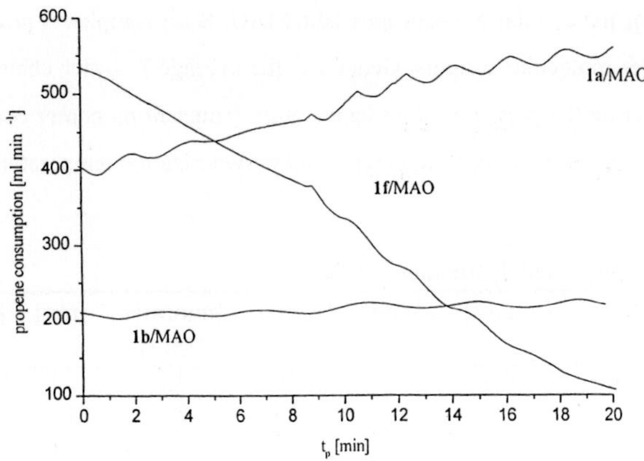

Figure 2. Propene consumption curves obtained with **1a,b,f**/MAO: **1a**/MAO: T_P = 50 °C, $[C_3]$ = 3.0 mol L^{-1}, n_{Zr} = 3.8 µmol and **1f**/MAO: T_P = 50 °C, $[C_3]$ = 3.0 mol L^{-1}, n_{Zr} = 10 µmol; **1b**/MAO: T_P = 30 °C, $[C_3]$ = 3.0 mol L^{-1}, n_{Zr} = 5.0 µmol.

The exchange of the ethylene bridge by a dimethylsilane unit (**1f-h**/MAO) leads to significantly different polymerization properties, especially to a deactivation, resulting in a rapid decline of the propene consumption during the course of the polymerization reaction (Figure 2, **1f**/MAO). Nevertheless, we suppose that also for the dimethylsilane bridged complexes **1f-h**/MAO a slow activation mechanism is present, which however is superimposed in the propene consumption curve by the fast decomposition of the active species.[12] This is in contrast to C_2-symmetric *ansa*-metallocenes that show higher activities for the Si-bridged species due to an enlarged bite-angle of the Cp-ligands in such complexes.[13]

Molecular Weight. In contrast, the Si-bridge strongly increases the polymer molecular weight. In that respect the 5,6-cyclopentyl-substituted complex **1f**/MAO (Table 1, entries 29,30) reached the highest molecular weights under comparable polymerization conditions. Interestingly, the 5,6-cyclopentyl group seems to have generally a beneficial influence on the molecular weight, despite its remote position within the complex relative to the active Zr(IV)-center (Table 1, **1a,f**). We have no genuine explanation for this surprising effect. However, experiments with deuterated propene monomers show, that the 5,6-alkyl substituents suppress effectively a chain end epimerization process (found to be responsible for the formation of stereoerrors in C_2-symmetric catalysts)[14] and hinder at the same time a subsequent chain termination reaction, leading to higher molecular weight products.[10b] Moreover we found a converse effect for the angular zirconocenes **1d,e**/MAO. Both complexes produced polymers with slightly lower molecular weights. Generally, the average polymer chain lengths can be increased by lowering the polymerization temperature (constant monomer concentration) and by enhancing the monomer concentration (constant polymerization temperature).

Table 2. Polypropene pentad distribution[a] (in %)

run	cat.	[mmmm]	[mmmr]	[rmmr]	[mmrr]	[mmrm]+ [rrmr]	[rrrr]	[rrrm]	[mrrm]
1	1a	47.8	16.9	1.0	18.6	3.2	0.8	2.0	9.3
2	1a	42.5	17.5	2.1	19.9	3.4	1.6	2.8	10.2
3	1a	30.3	17.7	2.7	21.3	5.9	4.6	6.6	10.9
4	1a	31.3	17.0	2.9	19.5	7.7	4.6	6.2	10.3
5	1a	27.8	17.7	2.7	22.0	6.6	5.3	7.4	10.5
7	1a	72.1	10.0	-	12.2	2.1	-	-	3.6
8	1a	63.9	12.7	0.3	14.2	1.0	0.3	0.7	6.6
9	1a	59.6	15.9	-	16.5	-	-	-	8.0
10	1a	64.0	14.2	-	12.9	1.2	-	0.9	5.8
11	1b	17.6	16.5	3.8	21.6	20.8	8.9	10.4	11.4
12	1b	13.4	15.6	5.4	22.5	10.3	14.0	11.8	7.0
13	1b	9.7	13.7	5.5	18.5	10.7	15.7	14.8	11.4
14	1c	98.0	2.0	-	-	-	-	-	-
15	1c	86.0	5.4	-	6.3	-	-	-	2.3
28	1f	61.6	15.0	-	16.3	-	-	-	7.1
29	1f	50.6	15.6	1.4	16.8	3.1	1.7	2.5	8.2
30	1f	44.0	16.3	1.8	17.8	4.1	3.0	3.9	9.1
31	1f	68.0	12.0	-	12.4	1.0	-	0.9	5.8
32	1f	67.4	12.6	-	13.0	0.9	-	0.5	5.5
33	1g	44.1	16.7	1.8	18.5	4.0	3.1	3.5	8.3
34	1g	39.3	17.5	2.3	20.6	3.8	3.7	4.8	8.0
35	1g	35.3	17.6	2.4	21.2	4.2	5.4	6.1	7.9
36	1g	56.2	16.4	-	18.0	2.1	-	1.6	5.8
37	1h	70.5	11.8	-	12.6	-	-	-	5.1
38	1h	62.2	12.9	1.1	12.0	2.5	0.9	1.7	5.9
39	1h	57.9	13.9	1.4	12.8	3.2	1.2	2.1	6.2
40	1h	71.7	11.7	-	11.9	-	-	-	4.7

[a] [mrmr] ≤ 1%.

Mechanisms of Stereoerror Formation

Stereoselectivity and temperature. The complexes **1a-h**/MAO gave isotactic polypropenes with variable amount of stereoerrors. In particular the [mmmm]-pentad concentration of the polymers produced with the different C_1-symmetric metallocene catalysts varies over a broad range (10% to 98%) (Tables 1,2; entries 14,15). Generally, the metallocenes **1d,e**/MAO, which show a 4,5-front substitution pattern, gave substantially higher isotactic polypropene products than the side-substituted catalyst **1a**/MAO (Table 2) or the unsubstituted species **1b**/MAO. This behavior can be attributed to the presence of a substituent in the 4-position of the indenyl moiety in **1d,e**, which favors a back-skip of the growing polymer chain before a new monomer inserts on the aspecific side of the complexes. The same effect, compared to the unsubstituted zirconocene **1b,** is also observed, but to a lower degree, for the complexes **1a,f**/MAO bearing the 5,6-cyclopentyl substitution on the indenyl fragment.

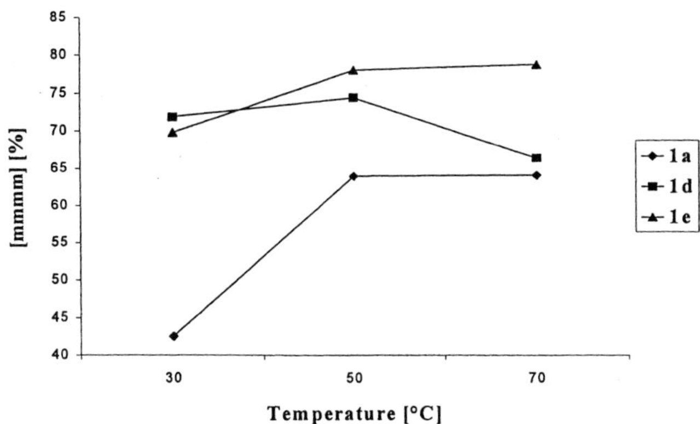

Figure 3. Plot of the propene stereoregularity versus the poylmerization temperature T_P for catalysts **1a,d,e**/MAO at constant monomer concentration ([C_3] = 1.1 mol L^{-1}).

Interestingly, there is no steady decline of the catalysts' stereoselectivities with increasing polymerization temperature (Figure 3) as typically observed for C_2-symmetric metallocenes.[13] The isotacticities of the products prepared with **1e**/MAO at a constant monomer concentration (e.g., 1.1 mol L^{-1}) increase with temperature from 70 (30 °C) to 79% [mmmm] pentads (70 °C). The zirconocene **1d**/MAO gives polypropenes of which the isotacticity goes through a maximum at 50 °C (74% [mmmm]), and the products of **1a**/MAO show an increase of the [mmmm] concentration from 30 °C (48%) to 50 °C (64%), after which it remains constant upon a further increase of the polymerization temperature to 70 °C.

The lowest isotacticities are observed for the non-substituted complex **1b**/MAO with the lowest steric demand reducing the rate of the growing polymer chain to undergo a chain backskip reaction (c.f. Scheme 1). This temperature dependence of the isotacticity on the polymerization temperature indicates that the stereoerror formation at higher temperatures is based on two different pathways. The stereoerror formation of asymmetric complexes originates predominantly from the kinetic competition between chain back-skip, that increasingly takes place with rising temperatures and monomer coordination at the aspecific side of the catalyst structure. As a result, we observe initially an increase of the [mmmm] pentads at elevated temperatures. The most conspicuous effect of the Si-bridged *ansa*-metallocenes **1f,g**/MAO is the distinct ascent of the isotacticity compared with their ethylene bridged counterparts **1a,b**/MAO. The series of propene polymerizations applying the 2-phenyl substituted complexes **1c**/MAO has produced polymers with the highest isotacticities (e.g **1c**: [mmmm] = 98%, Table 2, entry 14). Exchange of the C_2H_4-bridge by a Me_2Si-unit (**1h**) results in a strong decrease of the isotacticity of the polypropene products (e.g. Table 2, entries 37-40), while for both complexes the same influence of the monomer concentration on the stereoselectivity is observed (Table 2, **1c**: entries 14, 15; **1h**: entries 37-40), suggesting that a mechanism like the one depicted in Scheme 3 might still be active. Generally, 2-phenyl substitution leads to a strong decrease in polymerization activity, while the molecular weight is not essentially altered.

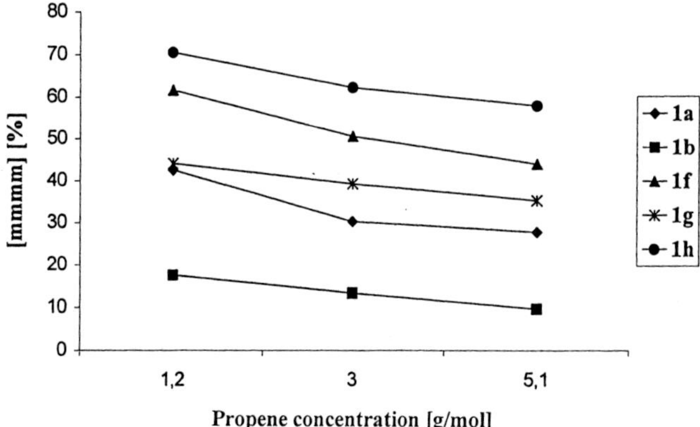

Propene concentration [g/mol]

Figure 4. Plot of the propene streoregularity versus the monomer concentration for catalysts **1a,b,f-h**/MAO at constant polymerization temperature (T_P = 30 °C).

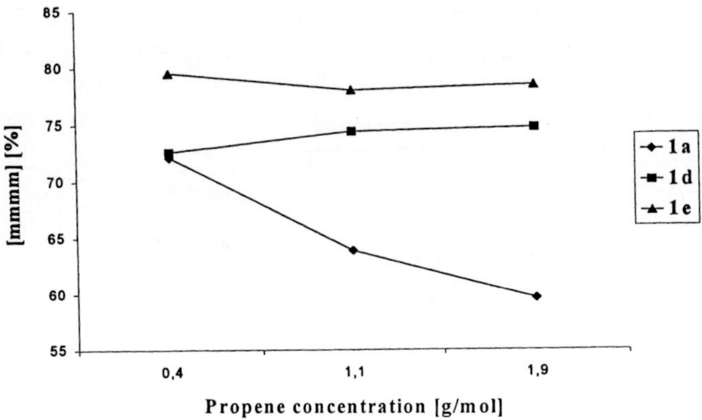

Figure 5. Plot of the propene streoregularity versus the monomer concentration for catalysts **1a,d,e**/MAO at constant polymerization temperature ($T_P = 50\ °C$).

Stereoselectivity and monomer concentration. The [mmmm] pentad content declines continuously with increasing monomer concentration (Figure 4 and Figure 5). This is a exceptional behavior that distinguishes our catalysts entirely from the polymerization properties of common C_2-symmetric metallocenes. The decrease of isotactic sequences is attributed to the existence of two different coordination sides in our type of asymmetric metallocene complexes. We assume that the two sides differ distinctly in their stereoselectivity toward the propene insertion reaction, a hypothesis, that was supported by Guerra et al. in a comprehensive theoretical study.[15] Isotactic sequences are formed by a reaction path, which involves repeated migratory insertion of the growing polymer chain into the propene monomer, coordinated between the sterically demanding fluorenyl-indenyl moieties, (Scheme 1, A → B) and a consecutive back-skip of the chain to the sterically less encumbered side (B → C). At low monomer concentrations, the chain moves back to the previous position (C → A), because the sterically less encumbered side of complex structure is energetically favored by the bulky polymer chain, resulting in consecutive insertion reactions of propene molecules at the sterically more hindered coordination side (A → B). Increasing the monomer concentration leads to a higher probability for monomer coordination and insertion at the less hindered side before a back-skip of the growing chain can occur.[15] Migratory insertion of the chain to the free side results in a non-selective insertion of the monomer (D → C). The next insertion from (A → B) is again stereoselective,[16] leading - after a back-skip of the polymer chain - to an isolated [mrrm]-error pentad.

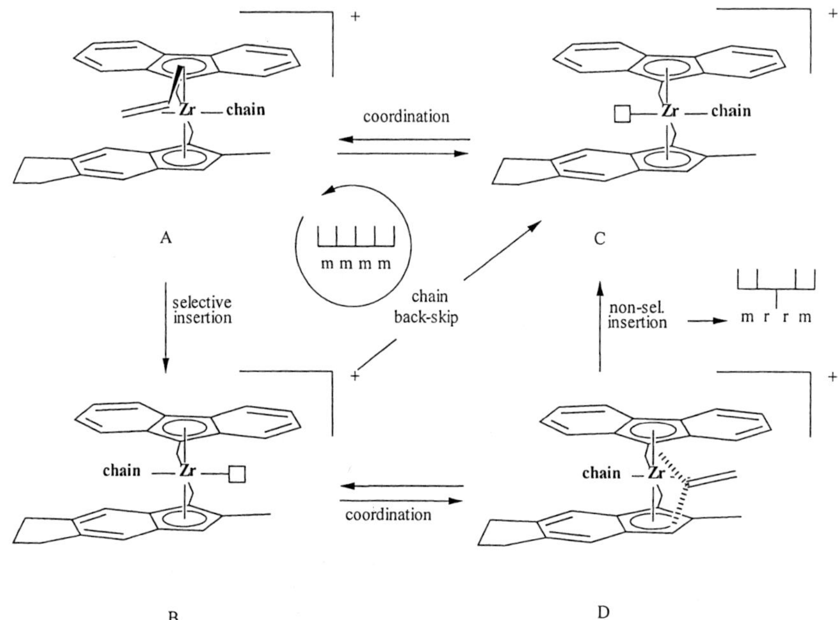

Scheme 1. Monomer concentration as a tool for the formation of stereoerrors in high molecular weight polypropenes with "dual-side" systems as presented in the present study.

At the same time as higher monomer concentrations result in a decline of the polymer isotacticity, all pentads being characteristic for an isolated stereoerror increase distinctively with monomer concentration. Figure 6 and Figure 7 show a fingerprint of the polymerization mechanism and point out that the proposed polymerization mechanism is independent of the bridging unit. For these catalysts the ratio of the rate of back-skip of the chain to the rate of monomer concentration at the free side can be understood to determine to a large extent the amount of isolated stereoerrors formed during the polymerization reaction. The amount of stereoerrors increases with the monomer concentration (c.f. Table 2, entry 1-6), because the rate of monomer coordination at the free side (B → D) depends on C_3-coordination while the rate of back-skip of the chain (B → C) is an intramolecular reaction.[17]

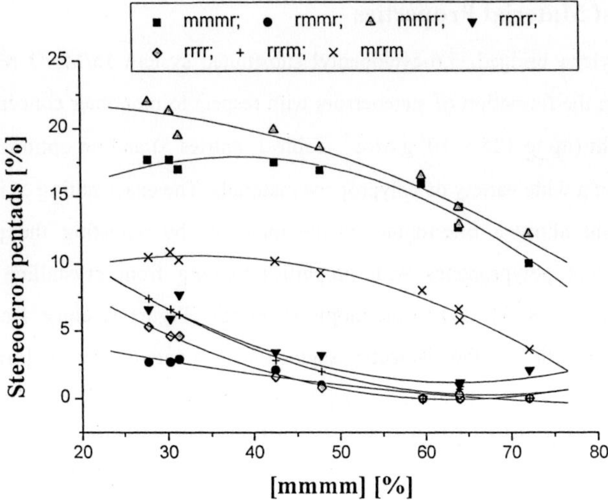

Figure 6. **1a**/MAO: Variation of stereoerrors in polypropenes with reduced isotacticity as a fingerprint for the polymerization mechanism.

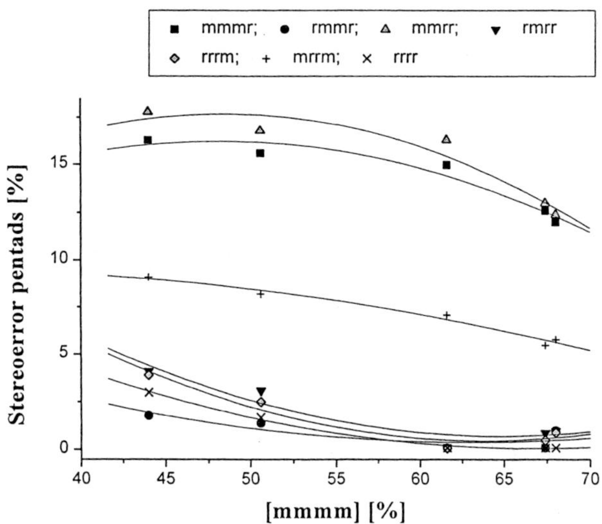

Figure 7. **1f**/MAO: Variation of stereoerrors in polypropenes with reduced isotacticity as a fingerprint for the polymerization mechanism.

The Design of Material Properties

The linear, ethylene bridged, 5,6-cyclopentyl substituted system **1a**/MAO reflects the best balance between the formation of stereoerrors with respect to monomer concentration, a high molecular weight (up to 125×10^3 g Mol^{-1}, Table 1, entries 5) and an optimized activity for the production of a wide variety of polypropene materials. The exact setting of isotacticity and molecular weight allow a fine-tuning of the materials by adjusting the polymerization conditions, so that polypropenes with properties ranging from crystalline thermoplastic (Figure 8, entries 8, 28, 31, 35) to thermoplastic elastic (Figure 8, entry 4 and 33) can be achieved. The length of the isotactic segments is - at increased [mmmm] pentad concentrations (higher temperatures, lower [C_3]) - long enough to afford a higher degree of crystallinity so that these materials display the behavior of tough thermoplastic plastomers with relatively high melting temperatures.[18] Interestingly, our simple homopolypropenes can compete with more complicated copolymers, like Engage (Figure 8, entry 43) and even Kraton (Figure 8, entry 42) which might point to a variety of applications.

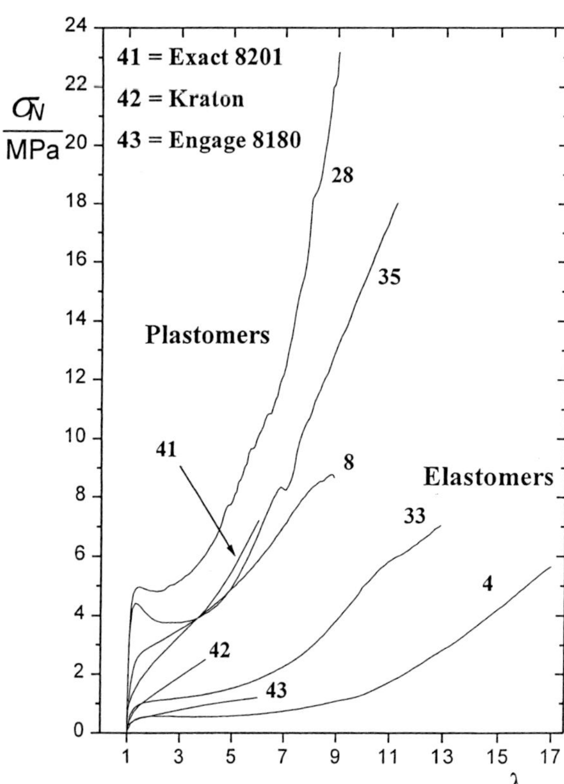

Figure 8. Typical stress-strain curves of selected polymers prepared with the zirconocenes **1a,f,h**/ MAO in comparison with the tensile behavior of commercial polymer samples (entries 41-43).

The maximal elongations, λ_{max}, of these partly crystalline polymers depend – besides the catalysts structure itself - on both the isotacticity and the molecular weight of the samples. Reduction of the isotacticity leads to the formation of highly elastic materials with λ_{max} = 17-20 times the original length.

It is important to point out that the elastic properties of the polypropene polymers obtained with our asymmetric complexes are not ascribed to chain entanglements as reported for high molecular weight, atactic polymers. For our homopolypropenes the interesting elastic property profile results obviously from a strengthening of the amorphous phase by a crystalline network, which can be adjusted by regulating the isotactic block length. The position and length of the elastic plateau of these materials can be tailored together with the initial force to stretch the polymers so that a continuous change of mechanical properties from elastic to thermoplastic can be achieved (Figure 8).

Conclusion

In the present study, we provided access to a wide variety of new polypropene materials resulting from a new family of asymmetric metallocene complexes. As a new tool, the C_1-symmetry of these "dual-side" catalysts allows to place single stereoerrors along an isotactic chain leading to a broad range of isotactic polypropene microstructures that can be accurately adjusted by variation of the monomer concentration and the polymerization temperature. A controlled tailoring of the properties of the polymer products from highly elastic to crystalline thermoplastic is enabled by varying the catalyst structure and the polymerization conditions. The interesting elastic property profile results from a reinforcement of the amorphous phase by a crystalline network, which is validated in preliminary scanning force microscopy studies giving a first insight into the crystallite size and the their distribution needed for the formation of a stable three dimensional network. Interestingly, the 5,6-cyclopentyl substitution on the indenyl moiety of the ethylene bridged species was found to play an important role in obtaining a good balance between optimized activity, increased molecular weight and a sufficient amount of stereoerrors for the design of thermoplastic elastomers. Detailed mechanistical studies support the 5,6-substitution to act as key position in the formation of stereoerrors, that originates predominantly from the kinetic competition between chain back-skip and monomer coordination at the aspecific side of the catalyst structure.[10] In contrast to most of the known catalyst systems for the production of elastic PP, of which structure variations are not possible or lead to non-elastic products, our complexes have a variable structure possessing unlimited possibilities for the design of new polypropenes. It will be interesting to see up to what extent this new family of high molecular weight, isotactic polypropenes can proceed in the resolute competition of commercialization.

Experimental Section

Complex synthesis: The syntheses of the complexes **1a-h** have been previously described in literature and were prepared according to these literature recipes.[10,19,20] Methylalumoxane and triisobutylaluminum were purchased from Witco and toluene for the polymerization reactions from Merck.

Propene Polymerization Reactions: The polymerization reactions were performed in a 1 L Büchi steel reactor at constant pressure and temperature. The autoclave was charged with 300 mL of toluene and with the desired amount of MAO. Subsequently, the polymerization temperature was adjusted, the reactor was charged with propene up to the desired partial pressure and the preactivated catalyst solution (Al:Zr = 100:1) was injected into the autoclave via a pressure burette. The monomer consumption was measured by the use of a calibrated gas flow meter (Bronkhorst F-111C-HA-33P), and the pressure was kept constant during the entire polymerization period (Bronkhorst pressure controller P-602C-EA-33P). Pressure, temperature and consumption of propene were monitored and recorded online. The polymerization reactions were stopped and treated as described above.

Polymer Analysis: ^{13}C NMR spectra were recorded on a Varian GEMINI 2000 spectrometer ($C_2D_2Cl_4$, 100 °C, 75 MHz, 10 mm probe), or on a Bruker AMX 500 spectrometer ($C_2D_2Cl_4$, 80 °C, 125 MHz, 5 mm probe) in the inverse gated decoupling mode with a 3 s pulse delay and a 45° pulse to attain conditions close to the maximum signal-to-noise ratio. The number of transients accumulated was between 5–15 K. The spectra were analyzed by known methods.[21] Molecular weights and molecular weight distributions were determined by gel permeation chromatography (GPC, Waters 150 C ALC, 135 °C in 1,2,4-trichlorobenzene) relative to polystyrene and polypropene standards. Mechanical measurements were performed on a Zwick 1445 tensile apparatus at room temperature with an extension rate of 10 mm min⁻¹. All samples were prepared under identical conditions by pressing the polymer melt (170 °C) in vacuo to 10- × 5- × 0.1-cm specimens, which were cooled to ambient temperature over 20 min.

Acknowledgment

Generous financial assistance is gratefully acknowledged from Deutsche Forschungsgemeinschaft (SFB 239, F 10), Fonds der Chemischen Industrie, Deutscher Akademischer Austauschdienst (DAAD) and Procter & Gamble. We also wish to thank Prof. Seppälä (Helsinki University of Technology) for the comprehensive support in polymer characterization.

[1] Wild, F. R. W. P.; Zsolnai, L.; Hüttner, G.; Brintzinger, H. H. *J. Organomet. Chem.* **1982**, 232, 233. (b) Wild, F. R. W. P.; Wasciucionek, M.; Hüttner, G.; Brintzinger, H. H. *J. Organomet. Chem.* **1985**, 288, 63. (c) Kaminsky, W.; Külper, K.; Brintzinger, H. H.; Wild, F. R. W. P. *Angew. Chem.* **1985**, 507, 97. (d) Ewen, J. A.; Jones, R. L.; Razavi, A.; Ferrara, J. D. *J. Am. Chem. Soc.* **1988**, 110, 6255. (e) Ewen, J. A.; Elder, M. J.; Jones, R. L.; Haspelagh, L.; Attwood, J. L.; Bott, S. G.; Robinson, K. *J. Am. Chem. Soc.* **1991**, 48/49, 253.
[2] (a) Spaleck, W.; Antberg, M.; Rohrmann, J.; Winter, A.; Bachmann, B.; Kiprof, P.; Behm, J.; Herrmann, W. A. *Angew. Chem. Int. Ed. Engl.* **1992**, 31, 1347. (b) Razavi, A.; Atwood, J. L. *J. Am. Chem. Soc.* **1993**, 115, 7529. (c) Spaleck, W.; Küber, F.; Winter, A.; Rohrmann, J.; Bachmann, B.; Antberg, M.; Dolle, V.; Paulus, E. F. *Organometallics* **1994**, 13, 954. (d) Giardello, M. A.; Eisen, M. S.; Stern, C. L.; Marks, T. J. *J. Am. Chem. Soc.* **1995**, 117, 12114. (c) Coates, G. W.; Waymouth, R. M. *Science* **1995**, 267, 217.
[3] Ewen, J. A.; Jones, R. L.; Razavi, A.; Ferrara, J. D. *J. Am. Chem. Soc.* **1988**, *110*, 6255. Ewen, J. A.; Elder, M. J.; Jones, R. L.; Haspeslagh, L.; Attwood J. L.; Bott, S. G.; Robinson, K. *Makromol. Chem., Macromol. Symp.* **1991**, *48/49*, 253.
[4] (a) Mallin, D. T.; Rausch, M. D.; Lin, Y.; Dong, S.; Chien, J. C. W. *J. Am. Chem. Soc.* **1990**, 112, 2030. (b) Chien, J. C. W.; Llinas, G. H.; Rausch, M. D.; Lin, Y.; Winter, H. H. *J. Am. Chem. Soc.* **1991**, 113, 8569. (c) Llinas, G. H.; Dong, S.-H.; Mallin, D. T.; Rausch, M. D.; Lin, Y.-G.; Winter, H. H.; Chien, J. C. W. *Macromolecules* **1992**, 25, 1242. (d) Llinas, G. H.; Day, R. O.; Rausch, M. D.; Chien, J. C. W. *Organometallics* **1993**, 12, 1283.
[5] (a) Gauthier, W. J.; Corrigan, J. F.; Nicholas, N. J.; Collins, S. *Macromolecules* **1995**, 28, 3771. (b) Gauthier, W. J.; Collins, S. *Macromolecules* **1995**, 28, 3778.
[6] Bravakis, A. M.; Bailey, L. E.; Pigeon, M.; Collins, S. *Macromolecules* **1998**, 31, 1000.
[7] Studies of Waymouth et al. on the unbridged bis(2-phenylindenyl)-zirconocene dichloride showed the possibility of producing thermoplastic polypropene elastomers. For references, see, e.g.: (a) Coates, G. W.; Waymouth, R. M. *Science* **1995**, *267*, 217-219. (b) Bruce, M. D.; Coates, G. W.; Hauptmann, E.; Waymouth, R. M.; Ziller, J. W. *J. Am. Chem. Soc.* **1997**, *119*, 11174-11182. (c) Tsvetkova, V. I.; Nedorezova, P. M.; Bravaya, N. M.; Savinov, D. V. Dubnikova, I. L.; Optov, V. A. *Polym. Sci., Ser. A* **1997**, *39* (3), 235-240 and literature cited there.
[8] The terms "δ-forward" and "λ-backward" refer to the conformation of twisted metallacycles; for definitions and closer discussion, cf.: (a) Corey, E. J.; Bailar, J. C., Jr. *J. Am. Chem. Soc.* **1959**, *81*, 2620. (b) Rieger, B.; Jany, G.; Fawzi, R.; Steimann, M. *Organometallics* **1994**, *13*, 647-653.
[9] According to theoretical investigations of Guerra and co-workers, this phenomenon could be attributed to energy differences (2-3 kcal mol -1) for propene coordination to the less hindered (favored) and the more highly substituted side (nonfavored), depending on the bridge conformation, cf.: Guerra, G.; Cavallo, L.; Moscardi, G.; Vacatello, M.; Corradini, P. *Macromolecules* **1996**, 29, 4834-4845.
[10] (a) Dietrich, U.; Hackmann, M.; Rieger, B.; Klinga, M.; Leskelä, M. *J. Am. Chem. Soc.* **1999**, 121, 4348. (b) Kukral, J.; Petri, L.; Feifel, T.; Troll, C.; Rieger, B. *Organometallics* **2000**, 19, 3767.
[11] Figure 2 shows only a small digest of all propene consumption curves; the given tendencies are valid for all complexes.
[12] First experiments of the corresponding dimethyl complex *rac*-[(9-η5-fluorenyl)-(2-methyl-1-η5-indenyl)dimethylsilane]zirconium dimethyl using borates as cocatalyst support these considerations.
[13] C.f. Brintzinger, H. H.; Fischer, D.; Mülhaupt, R.; Rieger, B.; Waymouth, R. *Angew. Chem.* **1995**, 107, 1255 and references therein.
[14] a) Leclerc, M. K.; Brintzinger, H. H. *J. Am. Chem. Soc.* **1996**, 118, 9024. b) Leclerc, M. K.; Brintzinger, H. H. *J. Am. Chem. Soc.* **1995**, 117, 1651. c) Busico, V.; Caporaso, L.; Cipullo, R.; Landriani, L.; Angelini, G.; Margonelli, A.; Segre, A. L. *J. Am. Chem. Soc.* **1996**, 118, 2105. d) Busico, V.; Brita, D.; Caporaso, L.; Cipullo, R.; Vacatello, M. *Macromolecules* **1997**, 30, 3971. e) Busico, V.; Cipullo, R.; Caporaso, L.; Angelini, G.; Segre, A. L. *J. Mol. Catal. A* **1998**, 128, 53. For isomerization via an allyl mechanism, see: f) Resconi, L.; Camurati, I.; Sudmeijer, O. *Topics in Catalysis* **1999**, 7, 145. g) Resconi, L. *J. Mol. Catal. A* **1999**, 146, 167.

[15] For theoretical calculations on the occurrence of back-skip of the growing chain in the case of asymmetric metallocenes, see: Guerra, G.; Cavallo, L.; Moscardi, G.; Vacatello, M.; Corradini, P. *Macromolecules* **1996**, 29, 4834.

[16] Monomer misinsertion from A → B after D → C will result in [mrmr]-error pentads, the concentration of which is however close to zero in polypropenes obtained with asymmetric catalysts of the type studied here.

[17] For a detailed discussion of the polymerization mechanism see: Dietrich, U.; Hackmann, M.; Rieger, B.; Klinga, M.; Leskelä, M. *J. Am. Chem. Soc.* **1999**, 121, 4348.

[18] The following melting temperatures were observed: entry 8: Tm = 96.0 °C; entry 28: 104.1 °C; entry 35: 72.6 °C.

[19] Emma, J. T.; Chien, J. W. C.; Rausch, M. D. *Organometallics* **1999**, 18, 1439.

[20] Rieger, B.; Jany, G.; Fawzi, R.; Steinmann, M. *Organometallics* **1994**, 13, 647.

[21] Busico, V.; Cipullo, R.; Corradini, P.; Landriani, L.; Vacatello, M.; Segre, A. L. *Macromolecules* **1995**, 28, 1887.

Terpyridines as Supramolecular Initiators for Living Polymerization Methods

*Marcel Heller[a] and Ulrich S. Schubert[b,c]**

[a] Lehrstuhl für Makromolekulare Stoffe, Technische Universität München, Lichtenbergstr. 4, 85747 Garching (Germany); [b] Center for NanoScience, Ludwig-Maximilians-Universität, Geschwister-Scholl-Platz 1, 80539 München, (Germany); [c] Laboratory of Macromolecular and Organic Chemistry (SMO), Eindhoven University of Technology, PO Box 513, 5600 MB Eindhoven (The Netherlands).

Summary: *Bis*(5,5"-*bis*(bromomethyl)-2,2':6',2"-terpyridine), *bis*-4'-(4-bromomethylphenyl)-2,2':6',2"-terpyridine and 4-hydroxymethyl-5',5"-dimethyl-2,2':6',2"-terpyridine metal complexes have been used as initiators for the living polymerization of 2-oxazolines and *L*-lactides. In both cases polymers with controlled molecular weights and narrow molecular weight distributions have been obtained. In-line diode array GPC measurements of iron(II) complexed poly(ethyloxazoline)s showed an unexpected absence of fragmentation. Viscosity experiments demonstrated the differences of the complexed and uncomplexed systems.

Introduction

Since bi- and terpyridines are very capable ligands to form stable transition metal complexes with interesting photochemical[1,2], electrochemical[3-5] or catalytic[4-8] properties, they have become a very active field of research. In order to combine these properties with tailor-made polymer materials, several paths have recently been described.[9-12] All these approaches result in more or less defined polymers with coordinative segments. One possible interesting application would be, e.g., the 'switching' of material properties (e.g. absorption/emission, viscosity, adhesion) due to the change of external conditions, such as, e.g., electrochemical, thermal or pH.

In order to obtain well-defined polymers with a determined number of chelating binding units, an effective concept is to use ligands or metal complexes as initiators for living polymerization procedures (e.g. polymerization of oxazolines, lactides or styrene). This concept has been well established for bipyridine systems by *Fraser* et al. as well as by our group.[13-20]

In this context, terpyridines are only little investigated. Due to their higher complexation constants[21] and a large variety of possible functionalizations they are nevertheless very interesting. In this article we will present new aspects of the use of terpyridine based initiators for the living polymerization of oxazolines and a new approach to poly(lactide)s. In the end an entrance into multifunctional terpyridine initiators is discussed.

Experimental

For UV/VIS spectroscopy, a Varian Cary 3 UV/Vis Spectrophotometer was used. In-line diode array GPC measurements were carried out with a Waters 996 Photodiode Array Detector attached to a Waters 717plus Autosampler GPC. The viscosity measurements were carried out with an Ubbelohde viscosimeter, capillary I. The concentration of the decomplexed polymer was $7.52*10^{-5}$ mol/10 ml. After each measurement, a calculated amount of $FeSO_4$ x 7 H_2O (~1 mg) was added to the solution (CH_3CN/CH_3OH; 1:1) until 100% of the polymer was fully complexed ($3.76*10^{-5}$ mol/10 ml) and no significant change in viscosity occurred.

Synthesis of terpyridine based supramolecular initiators for oxazoline polymerization: *Bis*(5,5"-*bis*(bromomethyl)-2,2':6',2"-terpyridine) iron(II) hexafluorophosphate (**3**) and *bis*-4'-(4-bromomethylphenyl)-2,2':6',2"-terpyridine iron(II) hexafluorphosphate (**1**) were synthesized according to published procedures (see ref[22]).

Poly(ethyloxazoline)s: Polymerizations were carried out in dry CH_3CN at 80 °C for 24 h according to ref[22,23]. After termination with piperidine and precipitation with diethyl ether, complexed poly(ethyloxazoline)s (**2, 4**) were obtained.

Decomplexation of supramolecular poly(ethyloxazoline)s: The iron(II) and cobalt(II) complexes of the polymer were refluxed in a mixture of CH_3CN with aqueous K_2CO_3 (1 molar). After precipitation, colorless polymers were obtained (see ref[22]).

4-Hydroxymethyl-5',5"-dimethyl-2,2':6',2"-terpyridine (5): Synthesized according to ref.[24]. Selected analytical data: ^{1}H NMR ($CDCl_3$, 300 MHz): δ (ppm) 2.37 (s, 6H), 3.63 (s, 1H), 4.81 (s, 2H), 7.60 (m, 2H), 8.32 (s, 2H), 8.42 (d, 2H, *J* = 7.98 Hz), 8.46 (s, 2H); MS (EI): m/z 291 (M^+, 90%), 290 (M^+- 1, 100%).

Poly(L-lactide)s (6): The polymerizations were carried out in silanized schlenk tubes under nitrogen. 4-Hydroxymethyl-5',5"-dimethyl-2,2':6',2"-terpyridine (**5**) dissolved in dry toluene (10 mg/6 ml) and the aluminium triethyl in toluene (0.33 equivalents per OH unit) was added to the schlenk tube. The reaction mixture was stirred for 30 min at room temperature. Afterwards the monomer (3S)-cis-3,6-dimethyl-1,4-dioxane-2,5-dione was added (the amount was calculated with regard to the desired molecular weights) and the mixture was stirred at 80 °C for 2 days. The reaction was quenched with methanol. After removal of the solvent, the residue was dissolved in a small quantity of dichloromethane and afterwards poured into cold methanol. The precipitated polymer was filtrated and dried *in vacuo.*

Results and Discussion

Poly(oxazoline)s from terpyridine based supramolecular initiators: Poly(oxazoline)s are polymers whose solubility (polarity-) properties can be adjusted easily by changing the substituents in the 2-position of the monomers. Due to the living character of the cationic polymerization method, both chain-end functionalization and block copolymerization can be realized.[25]

Halomethyl functionalized bipyridine metal complexes have been used by *Fraser* et al. and *Schubert* et al. to initiate the polymerization 2-oxazolines. Molecular weight control, low polydispersities (PDI) and linear [monomer]/[initiator] versus molecular weight plots proved the living character of the polymerization.[15,26] By this method, block copolymers have been synthesized.[13,27] In addition, the termination of the polymer chains was carried out with different functional groups.[28]

We extended this concept to the use of supramolecular initiators with terpyridine ligands. Ligands with *bis*bromomethyl groups at the outer pyridine rings (5,5"-positions)[23,29] (**1**) and with one benzylbromide group attached to the central pyridine ring (4'-position)[22,30] (**3**) were synthesized (Figure 1). Stille-type cross-coupling is a suitable method to prepare terpyridines with substituents at well-defined positions, starting from simple pyridine building blocks. After further functionalization reactions, supramolecular initiators, in particular with Fe^{II}, Co^{II} , Zn^{II} and Ni^{II} as the central metal ions were prepared. Oxazoline polymerization was carried out mainly with the Fe^{II}

complexes; however, single poly(oxazolines) were also prepared with Co^{II}, Zn^{II} and Ni^{II} complexes (Figure 1). Polymerizations with both types of ligands were shown to be living.

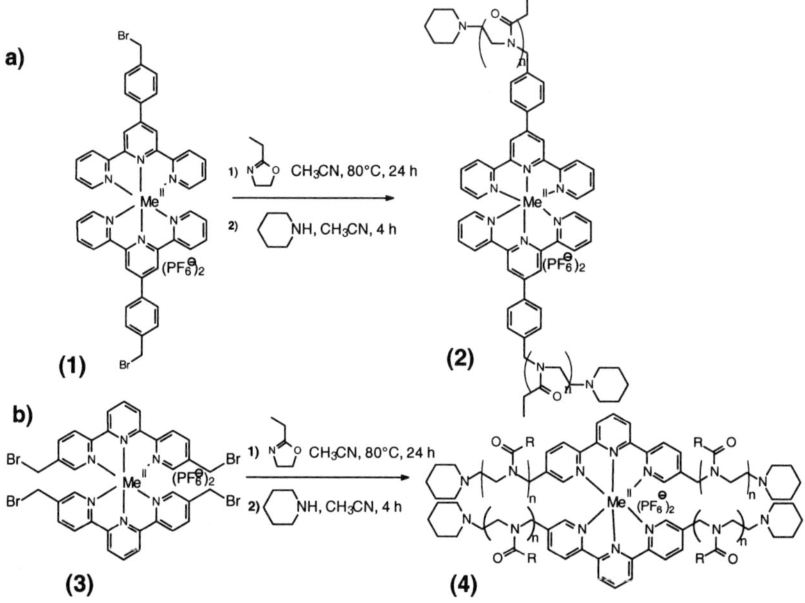

Figure. 1: Polymerization of 2-oxazolines utilizing supramolecular terpyridine based initiators.[22,23]

The polymer complexes can be cleaved by treatment with aqueous K_2CO_3. The incorporation of the metal ions in the polymers as well as the removal of the metal ion could be proven by UV/Vis spectroscopy. In order to examine the degree of reversibility of the complexation–decomplexation process, UV/Vis-titration experiments were carried out (Figure 2). It could be shown that recomplexation reaches up to 94%.[22] Titration of a decomplexed terpyridine terminated poly(ethyloxazoline) with $FeSO_4$ x 7 H_2O in CH_3CN/CH_3OH (1:1) was also performed in order to investigate the changes in viscosity behavior (Figure 3).

A typical 'ess' titration curve was obtained, showing a slight increase of viscosity at the point of equivalence. This could be expected in this case of a dimerization of the polymer. In contrast, the viscosity of star-like polymers – which can be obtained from (3) – should decrease compared to the open linear 'macroligands', caused by the smaller hydrodynamic radii of star-like polymers. Investigations concerning this subject are at present in progress.

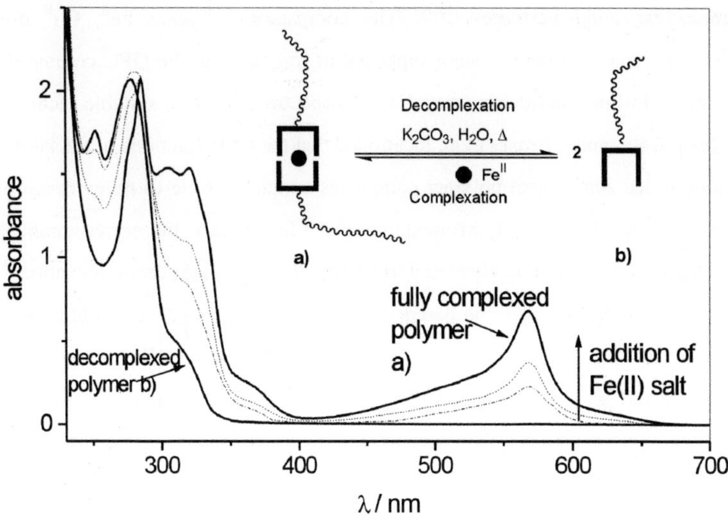

Figure 2: 'Supramolecular switching': UV/Vis-titration of a decomplexed poly(ethyl-oxazoline) solution with Fe^{II} salt.

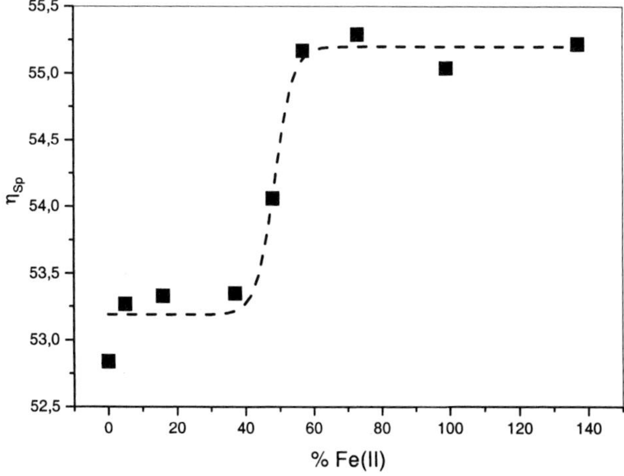

Figure 3: Changes of viscosity during titration of an uncomplexed poly(ethyloxazoline) $(7.52*10^{-3}$ mol/l, $M_n = 2660)$ with $FeSO_4 \times 7\ H_2O$ in CH_3CN/CH_3OH (1:1).

It is noteworthy that the molecular weights of the Fe^{II}-centered terpyridine-poly(oxazoline)s measured by GPC were only about half the expected values of the monomer/initiator ratios. After decomplexation, no significant change of the molecular weight occurred. The same phenomenon was observed at bipyridine based supramolecular poly(oxazoline)s.[15,26] The comparatively labile Fe^{II}, Co^{II} or Cu^{II} containing polymer complexes were supposed to fragment on the GPC column due to shear forces. In-line multidiode array UV/VIS-spectroscopy is a suitable technique to study this phenomenon. *Fraser* et al. mentioned that the GPC fractions of the labile Fe^{II}-bipyridine based star-shaped polymer complexes contained little to no evidence of the red violet Fe^{II} *tris*(bipyridine) chromophores.[19,27] In contrast, the corresponding inert Ru^{II} compounds were completely eluted from the column without decomplexation.[18]

In the case of iron(II)-terpyridine based poly(ethyloxazoline)s (**2**), we could show that the complexes did not fragment on the GPC column. The whole UV/Vis-spectra of the iron^{II} complexed polymer including the typical charge transfer band at 580 nm was detected by diode array UV/Vis detection (Figure 4). However, we cannot yet answer the question why the detected molecular weights are much lower than calculated.

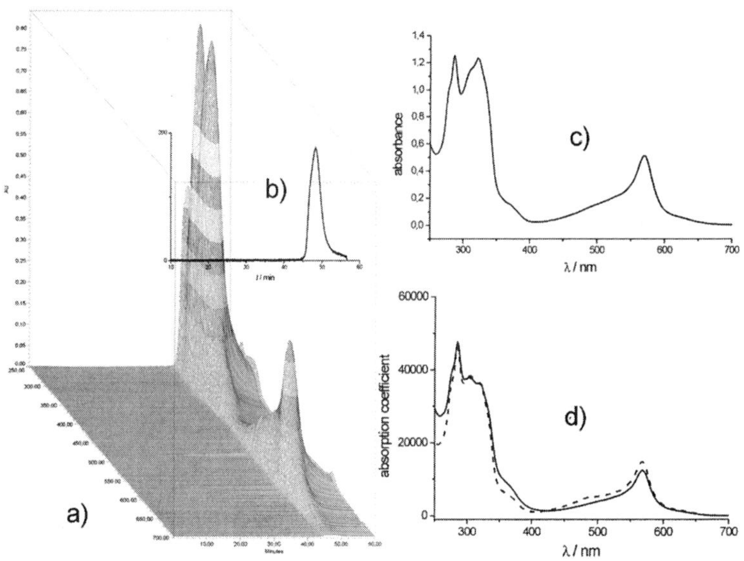

Figure 4: a) GPC-in-line diode array spectroscopy of a supramolecular Fe^{II}-poly(ethyloxazoline) (**2**) (M_n = 3370, M_w = 5200, PDI = 1.25). The x-axis represents the typical GPC curve, which is again shown in b) (RI detector). Along the y-axis the UV/Vis spectra is detected from 250-700 nm. c) shows a typical 2D-UV/Vis spectra which is the corresponding vertical section through the diode array curve. d) For comparison: UV/Vis spectra of the Fe^{II}-initiator complex (**1**) and a corresponding polymer (**2**).

Pol(lactides) from terpyridine based initiators: Poly(lactide)s are biodegradable polyesters which already found promising applications in medicine and tissue engineering.[31,32] Poly(lactide)s can be obtained via controlled coordinative ring-opening polymerization of lactides via aluminium alkoxides.[33,34] Hydroxymethyl substituted bipyridines were utilized recently as 'co-initiators' by *Schubert* et al.[35,36] and *Fraser* et al.[20,37] to yield bipyridine containing 'macroligands' or star-like polymer complexes. Meanwhile we extended this approach to terpyridine systems.

As co-initiator we utilized 4-hydroxymethyl-5',5"-dimethyl-2,2':6',2"-terpyridine (**5**) which can be prepared by reduction of 5',5"-dimethyl-2,2':6',2"-terpyridine-4-methylester (obtained using Stille-type cross-coupling reactions, see ref[24]). Polymerization of *L*-lactide yielded terpyridine end-capped poly(lactides) with controlled molar weight and very narrow polydispersities (Figure 5, Table 1). Further investigations are currently in progress.

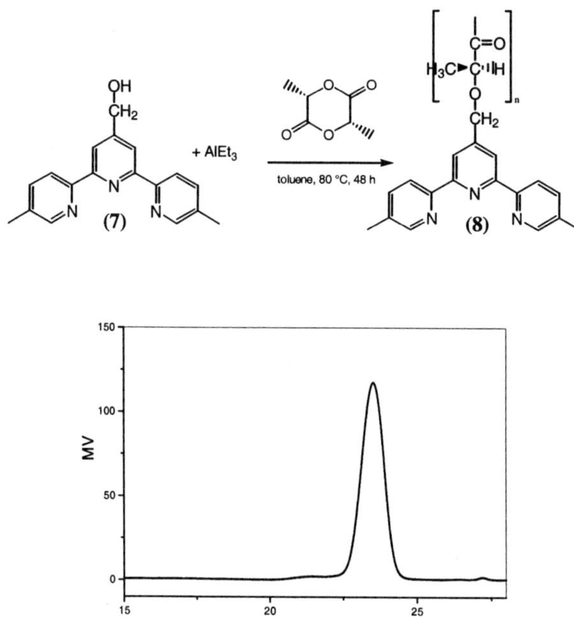

Figure 5: Polymerization of *L*-lactide with hydroxymethyl-functionalized terpyridine (**5**) as co-initiator.

Table 1: Various poly(L-lactide)s, polymerized utilizing co-initiator (**5**) (GPC data, polystyrene standards).

[M]/[I]	$\overline{\mathbf{M}_n}$ (RI)	$\overline{\mathbf{M}_w}$ (RI)	$\overline{\mathbf{M}_w}/\mathbf{M}_n$	calculated masses [g/mol]
16	3020	3250	1.08	2600
30	5045	5450	1.08	4500
31	3840	6100	1.40	4660
32	7230	8100	1.12	4800
52	9240	9680	1.06	7700
147	17770	24690	1.39	21300
287	41790	45510	1.10	41500

Conclusions

We could demonstrate, that the concept of supramolecular initiators for living polymerization methods such as cationic polymerization of 2-oxazolines or coordinative polymerization of lactides can be easily transferred to terpyridine systems. Both of the obtained polymer types revealed narrow polydispersities and the molecular weight could be controlled. Central metal ions such as Fe^{II} (or Co^{II}) can be removed under basic conditions and can be introduced again almost quantitatively. Viscosity measurements on the titration of terpyridine end-capped poly(ethyloxazoline)s with Fe^{II} ions showed the increase of viscosity and the typical S-form-titration curve expected for the dimerization. Contrary to first assumptions, supramolecular Fe^{II}-terpyridine centered polymers do not fragment on GPC columns. There must be another reason for the finding of half the molecular weights of Fe^{II}-terpyridine-poly(ethyloxazoline)s compared to the molecular weight adjusted by [monomer]/[initiator] ratios. Further studies like light scattering experiments have to be carried out.

Outlook

The way to multifunctional supramolecular initiators: In this article we presented functionalized terpyridine initiators for the polymerization of oxazolines and lactides. The initiating groups and the resulting polymers were either at the outer pyridine rings

(**1**) or at the centered pyridine ring (**3**). A promising extension of this concept would be the preparation of 'multifunctional' initiators – initiators with at least two different functional groups in one molecule (Figure 6). This would enable the grafting of two or more different polymers from one center or additional surface functionalization. Experiments in this direction are performed at present.

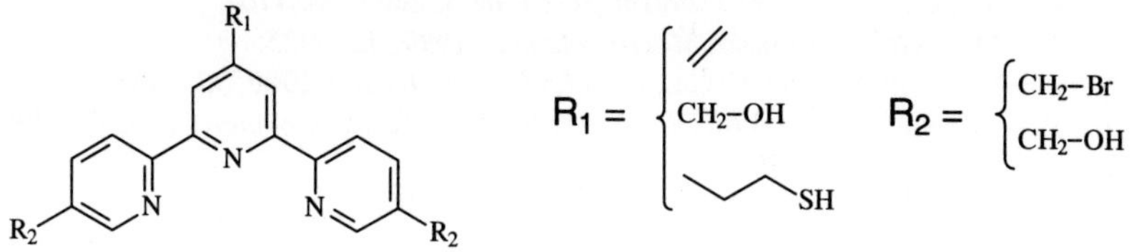

Figure 6: Schematically representation of a multifunctional supramolecular initiator.

Acknowledgements

The research was supported by the *Bayerisches Staatsministerium für Unterricht, Kultus, Wissenschaft und Kunst*, the *Fonds der Chemischen Industrie,* the *Deutsche Forschungsgemeinschaft (DFG)* and the *BAYER AG.* We thank O. Nuyken for his support.

References

[1]   H. Le Bozec, T. Renouard, *Eur. J. Inorg. Chem.* **2000**, 229.

[2]   G. Cross, *Nature* **1995**, *374*, 307.

[3]   J.P. Sauvage, J.P. Collin, J.C. Chambron, S. Guillerez, C. Coudret, V. Balzani, F. Barigelletti, L. De Cola, L. Flamigni, *Chem. Rev.* **1994**, *94*, 993.

[4]   C. Arana, M. Keshavarz, K.T. Potts, H.D. Abruna, *Inorg. Chim. Acta* **1994**, *225*, 285.

[5]   H. Nagao, T. Mizukawa, K. Tanaka, *Inorg. Chem.* **1994**, *33*, 3415.

[6]   E.B. Easton, P.G. Pickup, *Electrochem. Solid-State Lett.* **2000**, *3*, 359.

[7]   H. Nishiyama, T. Shimada, H. Itoh, H. Sugiyama, Y. Motoyama, *Chem. Commun.* **1997**, 1863.

[8]   J. Suh, S.H. Hong, *J. Am. Chem. Soc.* **1998**, *120*, 12545.

[9]   K. Hanabusa, A. Nakamura, T. Koyama, H. Shirai, *Makromol. Chem.* **1992**, *193*, 1309.

[10]   L.M. Dupray, M. Devenney, D.R. Striplin, T.J. Meyer, *J. Am. Chem. Soc.* **1997**, *119*, 10243.

[11]   W.Y. Ng, X. Gong, W.K. Chan, *Adv. Mater.* **1999**, *11*, 1165.

[12]   U.S. Schubert, C. Eschbaumer, *Macromol. Symp.* **2001**, *163*, 177.

[13]   U.S. Schubert, O. Nuyken, G. Hochwimmer, *Design. Monomer Polym.* **2000**, *3*, 245.

[14]   U.S. Schubert, O. Nuyken, G. Hochwimmer, *Polym. Mater. Sci. Eng.* **1998**, *79*, 11.

[15]   G. Hochwimmer, O. Nuyken, U.S. Schubert, *Macromol. Rapid Commun.* **1998**, *19*, 309.

[16]   U.S. Schubert, G. Hochwimmer, *Polym. Prepr.* **1999**, *40*, 340.

[17]   C.L. Fraser, A.P. Smith, *J. Polym. Sci., Part A.* **2000**, *38*, 4704.

[18]   J.E. McAlvin, C.L. Fraser, *Macromolecules* **1999**, *32*, 6925.

[19]   J.E. McAlvin, S.B. Scott, C.L. Fraser, *Macromolecules* **2000**, *33*, 6953.

[20]   P.S. Corbin, M.P. Webb, J.E. McAlvin, C.L. Fraser, *Biomacromolecules* **2001**, *2*, 223.

[21]   J. Prasad, N.C. Peterson, *Inorg. Chem.* **1969**, *8*, 1622.

[22]   U.S. Schubert, G. Hochwimmer, M. Heller, in *"Synthetic Macromolecules with Higher Structural Order, I."*. e. Khan, Ed., ACS Symp. Ser. **2001**, in press.

[23]   U.S. Schubert, C. Eschbaumer, O. Nuyken, G. Hochwimmer, *J. Incl. Phenom.* **1999**, *32*, 23.

[24]   M. Heller, U.S. Schubert, in preparation.

[25]   K. Aoi, M. Okada, *Prog. Polym. Sci.* **1996**, *21*, 151.

[26]   J.J.S. Lamba, C.L. Fraser, *J. Am. Chem. Soc.* **1997**, *119*, 1801.

[27]   J.E. McAlvin, C.L. Fraser, *Macromolecules* **1999**, *32*, 1341.

[28]   U.S. Schubert, O. Nuyken, G. Hochwimmer, *J. Macromol. Sci., Pure Appl. Chem.* **2000**, *A37*, 645.

[29]   U.S. Schubert, C. Eschbaumer, G. Hochwimmer, *Synthesis* **1999**, 779.

[30]   U.S. Schubert, M. Heller, G. Hochwimmer, *Polym. Prepr.* **2000**, *41*, 932.

[31]   R. Jain, N.H. Shah, A.W. Malick, C.T. Rhodes, *Drug Dev. Ind. Pharm.* **1998**, *24*, 703.

[32]   R. Langer, *Acc. Chem. Res.* **2000**, *33*, 94.

[33]   H.R. Kricheldorf, M. Berl, N. Scharnagl, *Macromolecules* **1988**, *21*, 286.

[34]   H.R. Kricheldorf, D.-O. Damrau, *J. Macromol. Sci., Pure Appl. Chem.* **1998**, *35*, 1875.

[35]   U.S. Schubert, G. Hochwimmer, *Polym. Prepr.* **2000**, *41*, 433.

[36]   U.S. Schubert, G. Hochwimmer, *Macromol. Rapid Commun.* **2001**, *22*, 274.

[37]   P.S. Corbin, J.E. McAlvin, M.P. Webb, S. Shenoy, C.L. Fraser, *Polym. Prepr.* **2000**, *41*, 1199.

Ultra-Thin Layers of Phosphorylated Cellulose Derivatives on Metal Surfaces

Evelin Jaehne[a], Thomas Kowalik[a], Hans-Juergen P. Adler[a,], Andreas Plagge[b], Martin Stratmann[b]*

[a]Dresden University of Technology, Institute of Macromolecular Chemistry and Textile Chemistry, Mommsenstr. 13, D-01062 Dresden, Germany

[b]Max-Planck-Institute of Iron Research, Duesseldorf, Max-Planck-Str. 1, D-40237 Düsseldorf, Germany

Summary: Cellulose as natural and non-toxic material is very interesting for biological applications. Its poor solubility in organic solvents can be improved by introduction of long alkyl chains. Phosphate-substituted cellulose derivatives form ultra-thin layers on several metal surfaces [1, 2]. The layers were applied on the surfaces via dip coating from dilute solutions and characterised by contact angle measurements. Initial corrosion tests were performed. These polymers were applied as adhesion promoters for steel-, titanium- and hydroxy apatite surfaces. Medical implants were pre-treated by UV irradiation, coated with the cellulose derivatives and activated by hot water treatment. The adhesion of the implants surpassed that of conventionally used PMMA systems.

Introduction

The present work deals with the application of biocompatible cellulose derivatives on implant materials in medicine.

Cellulose shows some very interesting properties: On one hand it is non-toxic, has a good thermal stability and due to its hydroxyl groups a high polyfunctionality. On the other hand cellulose possesses poor solubility in organic solvents and low thermo plasticity. And, natural celluloses do not form stable and ordered layers on metal surfaces.

The poor solubility of cellulose in technical solvents can be improved by introduction of alkyl chains. These polymers consist then of rigid main chains and liquid-crystalline side chains (see figure 1)

Wegner et al. investigated Langmuir-Blodgett layers formed from so-called "hairy-rod" polymers on the basis of cellulose derivatives, which possess hydrophobic side chains with a length of 4 –5 methylene groups.

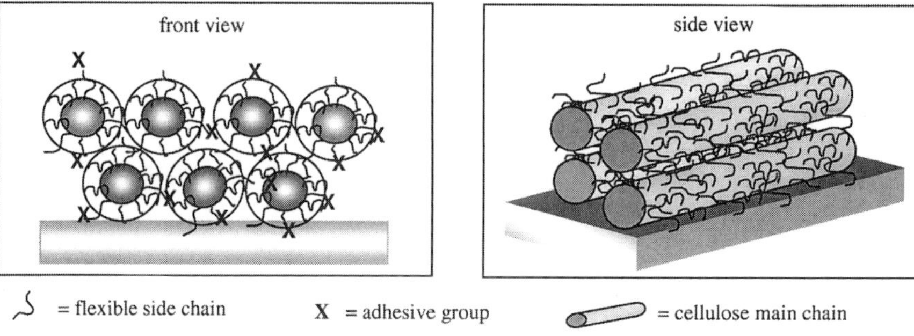

Figure 1: Scheme of "hairy rods" in side and front view

These "hairy-rod" cellulose polymers are able to form mono- and multilayers on hydrophobic surfaces [3-6]. It was possible to get regenerated cellulose with high biocompatibility by hydrolysis of cellulose silylesters [3].

Due to the easier handling required for the self-assembling technique [7] it was our intention to synthesise functionalised cellulose derivatives that are soluble in water or ethanol [8]. The functionality can be reached by introduction of functional groups into the terminal positions of the side chains. A synthetic route was developed to phosphorylate water-soluble hydroxypropyl cellulose [9-11].

The introduction of phosphate groups into the cellulose was the premiss for developing a system that was able to solve several adhesion problems in medicine. Main problems are that implants often loose the support to the bone and that tissue inflammation occurs. This can be solved by using modified layers between implant and bone to improve adhesion and minimize inflammation by incorporation of antibiosis. Implants of steel or titanium have been bond via the modified cellulose derivative to the bone (hydroxy apatite).

Thin films from phosphorylate hydroxypropyl cellulose were formed using the self-assembling technique on aluminium and titanium. We could observe a good adhesion on these substrates caused by the effect of the phosphoric acid groups [1, 2].

The characterisation of the films was carried out with contact angle measurements. Adhesion promoters for implant materials in medicine should not cause any corrosion. Therefore, initial investigations on the corrosion inhibition capability on aluminium surfaces were carried out using the constant humidity climate test (DIN 50017) for pre-treated metal plates. A second test was carried out in which the metal plates (Al, steel, Ti) were pre-treated with the cellulose derivatives and then coated with a clearcoat that was then damaged to assess the infiltration. Further investigations were made to stabilise the films of phosphorylated cellulose derivatives with a matrix of cross-linkable cellulose derivatives. For analytical investigations the composite was cross-linked by UV-radiation. This procedure is not usable for an application in the human body. Therefore, a new method was developed for cross-linking the composite [13].

Experimental part

All reagents and solvents were distilled and dried before using. Hydroxypropyl cellulose (Aldrich, MS = 4,8 (NMR), M_w ~ 100.000 g/mol,) was used without further purification. The synthesis' of hydroxypropyl-2-phosphatepropyl cellulose **1** [12] and hydroxypropyl-2-cinnamoylpropyl cellulose **2** [12] were described elsewhere [1].

Al substrates (AlMg1) of technical grade were provided by Chemetall GmbH, Frankfurt. The pre-treatment of the surface was published recently [12]. Ti (TiAL6V4) and steel were used after ultrasonically cleaning in acetone. Si substrates were ultrasonically cleaned in acetone for 15 minutes, immersed for one hour in a solution of H_2O/H_2O_2 (30 %)/NH_3 (25 %), ratio 5:1:1 at 80 °C, rinsed with water and dried under a stream of nitrogen.

Contact angle measurements

A K12 processor tensiometer from Krüss GmbH was used to measure the contact angles of the coated Al substrates with known geometry by the Wilhelmy method. Advancing and receding contact angles were calculated.

Constant humidity climate test [14]

This test was performed according to DIN 50017. The tests serve to clarify the corrosion protection of pre-treated aluminium plates under the influence of constant humidity at 40 °C for 96 h.

Industrial testing with clear coat top layer on aluminium substrates

The aluminium substrates were coated first with the thin cellulose derivative films and then with a commercial acrylate clearcoat (Weber + Wirth dried for 24 h in air). The coatings were mechanically damaged, activated with HCl-vapour for 1h and then exposed to the influence of constant humidity at 40 °C for 1000h. The results are described in mm infiltration.

Coating

The cleaned substrates were dip-coated in solutions at various concentrations (0,1-3 %) and for different dipping times (1-24 h), rinsed with water and dried in vacuum or a stream of nitrogen.

Results and discussion

One of our aims was to use the self-assembly process for the coating of the several substrates with cellulose derivatives. Therefore, water and alcohol-soluble cellulose derivatives have been developed [1].

Hydroxypropyl-2-phosphatepropyl-cellulose (**1**) was synthesised via phosphorylation of hydroxy propyl cellulose with polyphosphoric acid. The degree of substitution was 1,3 (see scheme 1).

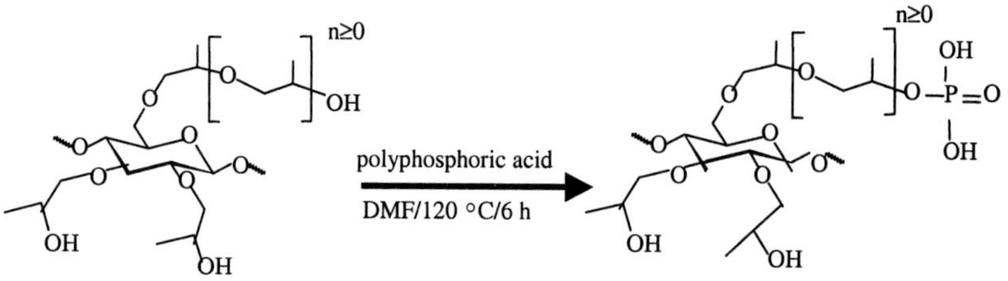

$DS_P = 1,3$

Hydroxypropyl-2-phosphatepropyl-cellulose (**1**)

Scheme 1: Synthesis of water-soluble hydroxypropyl-2-phosphatepropyl-cellulose (**1**)

The second component was a soluble cross-linkable cellulose derivative. Hydroxypropyl-2-cinnamoylpropyl cellulose (2) was prepared by introduction of cinnamoyl group into hydroxy propyl cellulose (see scheme 2).

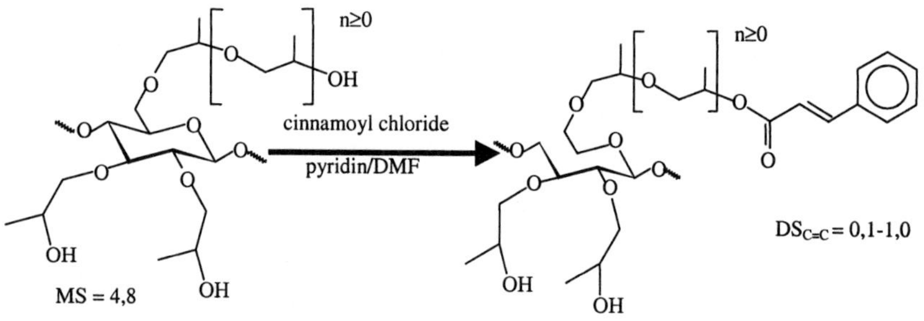

$DS_{C=C} = 0,1\text{-}1,0$

Hydroxypropyl-2-cinnamoylpropyl cellulose (2)

Scheme 2: Synthesis of soluble cross-linkable hydroxypropyl-2-cinnamoylpropyl cellulose (2)

The degree of substitution could be varied between 0,1 and 1,0.

The adhesive and corrosion resistant effects of the cellulose derivative layers strongly depend on the layer thickness. The layer thickness was measured with ellipsometry. For these measurements it was necessary to apply the ultra-thin layers on silicon substrates because of their smooth surface (see table 1).

Solution concentration	0,1%	2%	3%
Layer thickness	1• • • • • • nm	5,9• • • • • nm	Could not be determined, too inhomogeneous

Table 1: Layer thickness of **1** on Al in dependence on the solution concentration

The thin layers of hydroxypropyl-2-phosphatepropyl cellulose **1** from dilute solutions have a thickness of one nanometer. The layer thickness got at higher concentrations indicates that multilayers have been formed.

The layer of the cellulose **1** (from 0,1 % concentrated solutions) gave constant contact angles. The contact angle measurements for the layer of **1** (from more highly concentrated solutions) were not constant (see figure 2). The layer probably either dissolves or adsorbs water. The measurements for cellulose **2** could not be clearly interpreted. It possibly demonstrates the same behaviour as cellulose **1** from highly concentrated solutions.

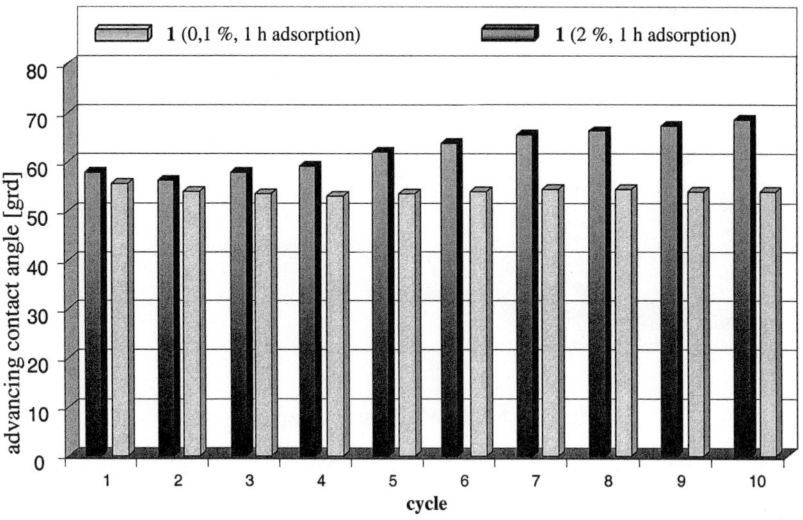

Figure 2: Stability measurements with dynamic contact angle analysis of **1** on Al

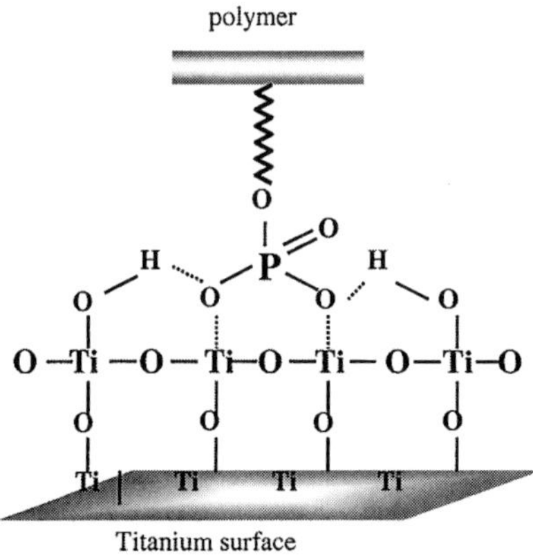

The adsorption process of phosphate groups on oxide surfaces can be considered as acid-base reactions. Figure 3 shows the model of adsorbed phosphate cellulose on the titanium surface.

Figure 3: Model for binding of phosphate cellulose with hydroxyl groups on titanium surface

Cross-linking

Cross-linking of the cinnamoyl groups occurred via a fast isomerisation to the cis-form followed by the formation of the cyclo-form via a slower reaction. A concentration of 10-20 % cross-linking groups was necessary for a stable cross-linking of the layer [2,13]. Figure 4 shows a model how the phosphorylated cellulose can be stabilised in an adhesive composite by the UV-crosslinked cellulose.

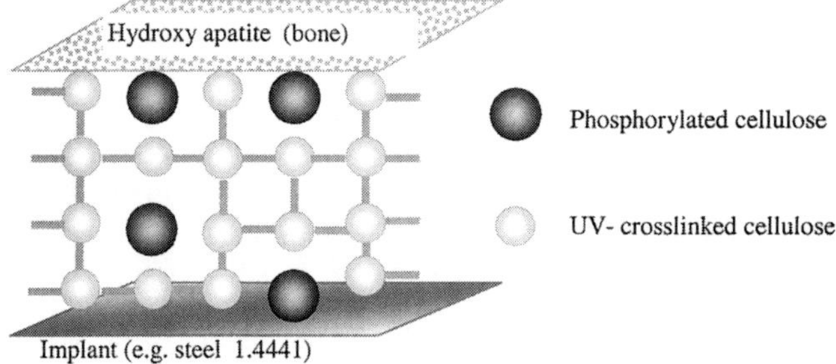

Figure 4: Model of adhesive composite

We investigated the cross-linking of the layer using UV-spectroscopy on a layer of **2** on quartz glass as shown in figure 5.

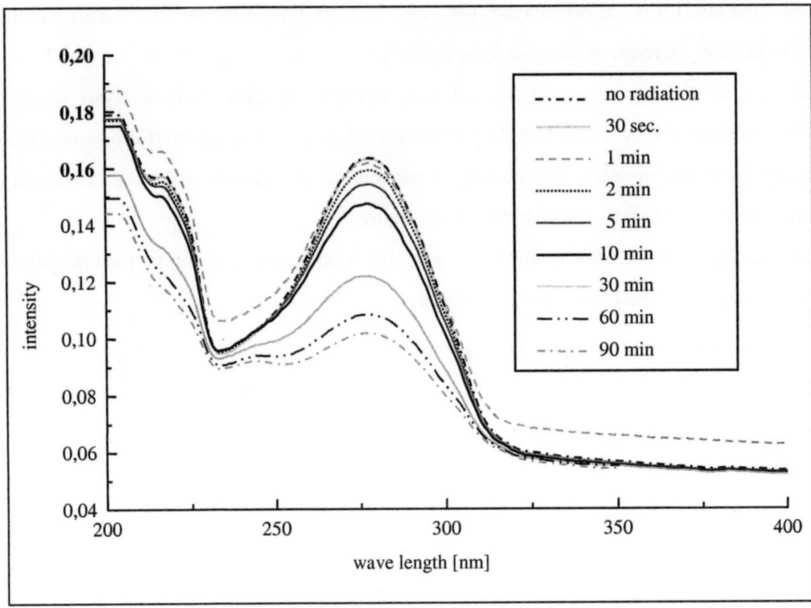

Figure 5: UV spectra of **2** on quartz glass in dependence on the radiation time

Reiser et al.[15] reported a method to calculate the amount of *trans*-, *cis*- and *cyclo*-form in the layer. The calculated amounts are shown in figure 6.

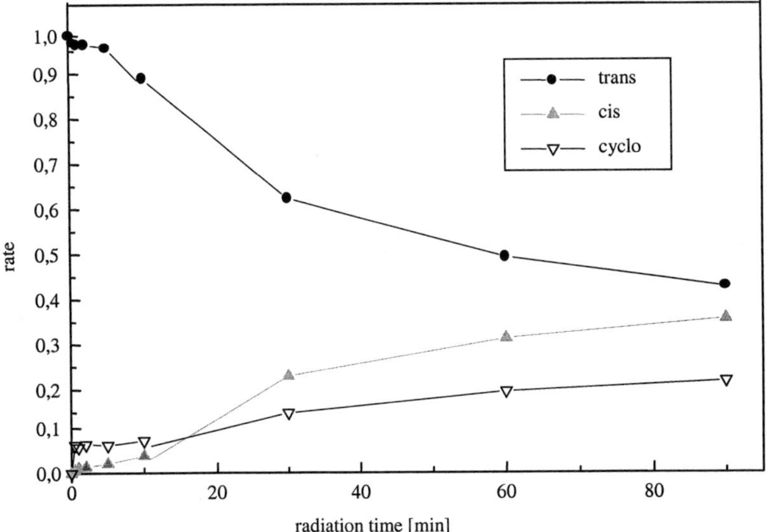

Figure 6: Rate of trans-, cis- and cyclo-form in the layer of **2** on quartz glass

After an irradiation time of 30 minutes the layer is cross-linked to 14 %, a longer radiation time only gave an increase of cross-linking to 20 %.

The stability of the layers was measured with dynamic contact angle analysis using the Wilhelmy method. Cyclic contact angle measurements give an indication of the layer stability (ten cycles for every sample). The coating is resistant to the solvent when the contact angle does not change. The results are summarised in figure 7.

The contact angles increased after the first cycle due to the change of the surface quality (dry and wet surface), after that they remained stable.

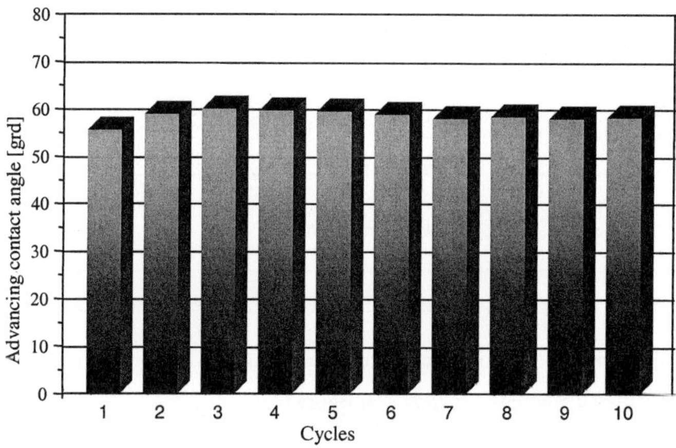

Figure 7: Cyclic contact angles of UV-cross-linked cellulose composite with high phosphate content

Corrosion tests

Fur using these cellulose derivatives in medical implant technology they have to fulfil two functions – first an adhesive and second a corrosion resistant function. Therefore, the corrosion inhibition effect of the two cellulose derivatives has been tested by constant humidity climate test and filiform test.

The constant humidity climate test showed that only the layer of **1** (0,1 % solution) provided good protection after 24 hours in comparison with a non-coated substrate. The corrosion inhibition effect is caused due to the phosphate groups. But after 96 hours there was no further corrosion inhibition. The layers of **1** (1-3% solution) and **2** show no inhibition potential (see figure 8).

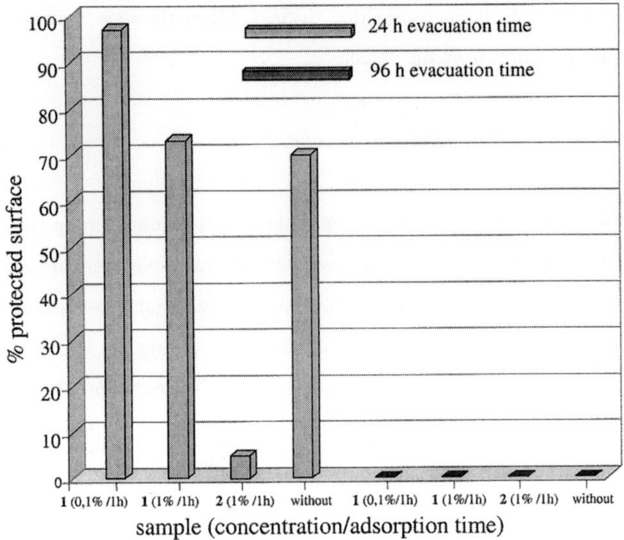

Figure 8: Constant humidity climate test of **1** and **2** on Al

The so-called filiform test is performed to investigate the tendency of filiform corrosion on scratched metal plates that were treated with the cellulose derivatives and a clear coat. The filiform corrosion tests show that the corrosion protecting effect can be observed, but it is strongly depended on the layer thickness of the cellulose derivative. The infiltration remained below 0.5 mm. The comparison of two different layer thickness' is shown in figure 9.

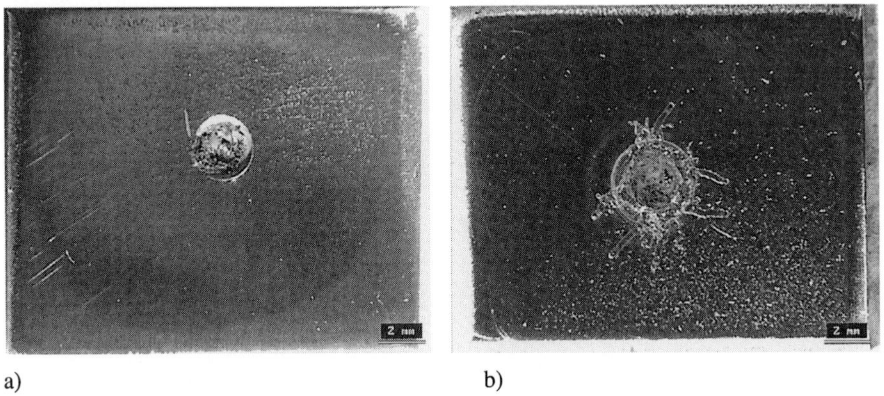

a)   b)

Figure 9: Filiform corrosion test of coated steel substrates

In picture a) steel 37 was coated by dipping with a very thin layer of (**1**) from 0,1% solution and a following protecting coating. After 120 h exposure time there is no corrosion outside of the defect. In picture b) a thicker cellulose layer was applied under the clear coat by spin coating. In this case after 120 h exposure time distinct filaments were visible formed by infiltration of corrosion products from the defect.

New method for application under body-like conditions [2, 13]

The use of the cellulose composite as bone cement replacement requires strong fixation to the substrate under mild conditions. Therefore, a new method was developed for activation of adhesion between bone and implant. First the implant material was treated with an ethanol/water solution of the two cellulose derivatives. After drying the implant is covered with a firm cellulose film. Then this layer is cross-linked by UV irradiation. By adding water the film develops its adhesive effect. Therefore, the implant is shortly dipped into hot water (95°C). The film begins to swell without heating the metal and develops its adhesive strength. Then the composite consisting of implant, adhesive cellulose derivative and bone is fixed by pressure.

The strength of the composite was investigated with the tensile test (DIN 4624).

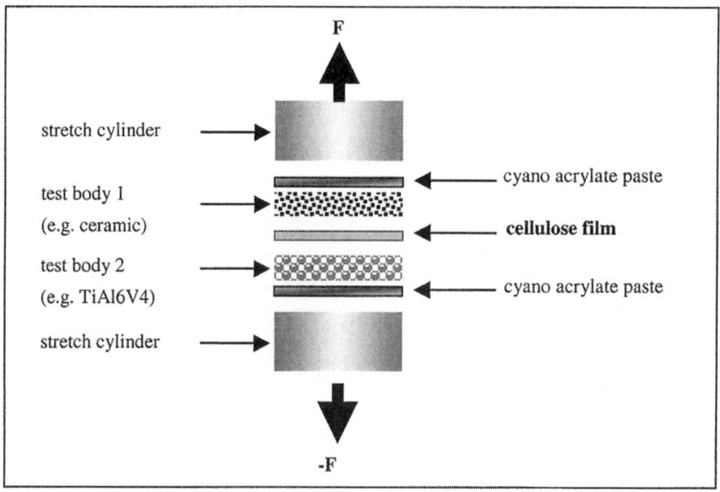

Figure 10: Set-up for measurement of adhesion effect by DIN 4624

In figure 10 the set-up for this measurement is shown. TiAl6V4, 1.4441 steel (contains 16-18% Cr, 10.14% Ni and 2-3% Mo) and hydroxy apatite were used as substrates. The substrates were coated with the cellulose derivatives, dried and then UV cross-linked. As cellulose derivatives were used different mixtures of the components: pure cinnamate cellulose, CP 91 containing 10% and CP 21 30% phosphate cellulose and pure cellulose phosphate. After activation in hot water the composite is pressed and stored 24h in air. For the tensile test a speed of 10 mm/min was chosen. The results are demonstrated in figure 11.

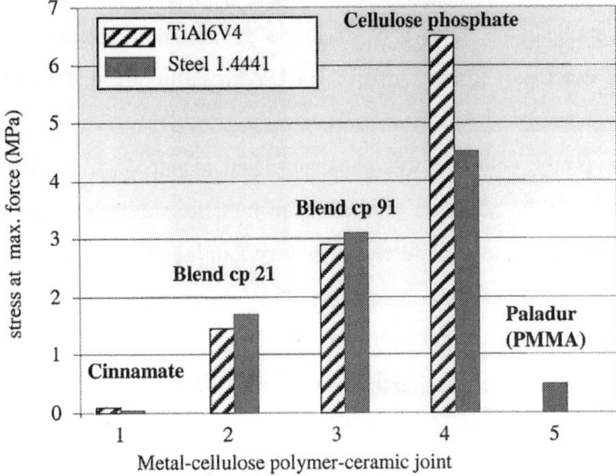

Figure 11: Comparison of different adhesive composites

From figure 11 one can conclude that any cellulose composite shows better results than the commercial product Paladur (PMMA). There are some differences between steel and titanium; the stress on the titanium substrate is 33% higher compared to steel. The adhesion between hydroxy apatite and cellulose phosphate is so strong that the break occurs in the ceramic (Fig.12).

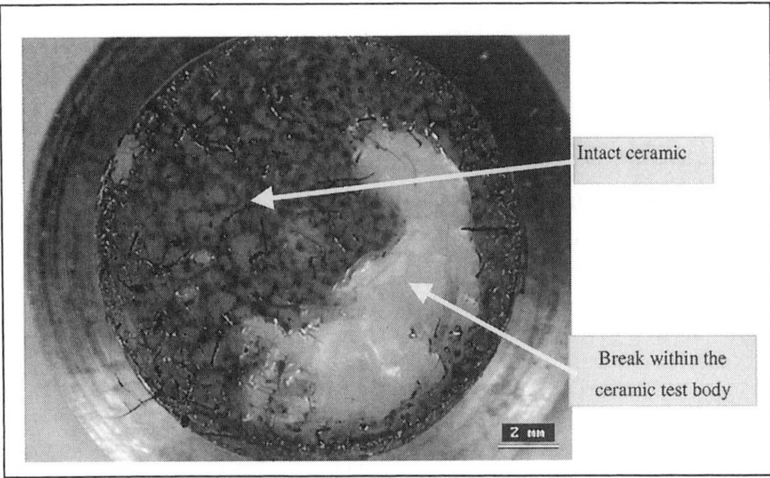

Figure 12: Front view on ceramic test body (cellulose phosphate adhesive) after tensile test

(DIN 4624)

Conclusions

Two-component adhesives from different cellulose derivatives have been developed for medical implant applications. One part functions as adhesion promoter for the several substrates and the other part forms a matrix via UV cross-linking that stabilises the whole composite.

Therefore, hydroxypropyl cellulose was phosphorylated to introduce functional groups into the cellulose increasing the adhesion promotion on metallic substrates. The phosphorylated cellulose derivatives were able to build closed and smooth layers on the metal surfaces with a preferred orientation of the cellulose main chain. They showed good adhesion promotion, but no significant improvement of the corrosion inhibition without a protecting coating on aluminium. The effect of adhesion promotion strongly depended on the concentration of the adsorption solution: dilute solutions are more effective than concentrated ones.

UV-cross linked cinnamoyl cellulose was synthesized to stabilize the phosphate cellulose in a matrix.

It was possible to show that thin layers of phosphated cellulose inhibited electrochemical corrosion reactions. These layers had to be protected due to the hydrophilicity of the cellulose derivatives by a top layer. A new technique has been developed for the application of these cellulose derivatives in medicine. The activation of the adhesive was carried out by the treatment of the pre-cross-linked cellulose layer on the implant with hot water as swelling agent. We could get stable composites on hydroxy apatite ceramic, titanium alloy (TiAl6V4) and steel (1.4441).

Acknowledgements: We are grateful to the Deutsche Forschungsgemeinschaft (DFG) for financial support through the foundation program "Cellulose and Cellulose Derivatives – Molecular and Supramolecular Structure Design".

References

[1]   T. Kowalik, H.-J. P. Adler, A. Plagge, M. Stratmann: Macromol. Chem. Phys. 2000, 201, 2064

[2]   A. Plagge, M. Stratmann, H.-J. Adler, T. Kowalik: Farbe & Lack *106*, 11 (2000) 48

[3]   M. Schaub, K. Mathauer, S. Schwiegk, P. A. Albouy, G. Wenz, G. Wegner: Thin Solid Films 1992, 210/211, 397

[4]   P. A. Albouy, M. Schaub, G. Wegner: Acta Polymerica 1994, 45, 210

[5]   G. D'Aprano, C. Henry, A. Godt, G. Wegner: Macromol. Chem. Phys. 1998,199, 2777

[6]   F. Löscher, T. Jaworek, T. Ruckstuhl, G. Wegner: Langmuir 1998, 14, 2786

[7]   T. Arndt, H. Schupp, W. Schrepp: Thin Solid Films 1989, 178,319

[8]   W. Wagenknecht, B. Philipp, I. Nehls, M. Schnabelrauch, D. Klemm, M. Hartmann: Acta Polym. 1991, 42, 213

[9]   D. Klemm: Acta Polym. 1997, 4, 277

[10]   M. Schaub, G. Wenz, G. Wegner, A. Stein, D. Klemm: Adv. Mater. 1993,5, 919

[11]   V. Buchholz, G. Wegner, S. Staume, L. Ödberg: Adv. Mater. 1996, 8, 399

[12]   T. Kowalik, Ph.D. Thesis, TU Dresden, 1999

[13]   A. Plagge, PhD Thesis, Uni Erlangen, 2000

[14]   I. Maege, E. Jaehne, A. Henke, H.-J. Adler, C.Jung, C. Bram, M. Stratmann: Prog. Org. Coat. 1998, 34, 1-12

[15]   P. L. Egerton, E. Pitts, A. Reiser, Macromolecules 1981, 14, 95

Fabrication of Carbon/Silica Hybrid Materials Using Cationic Polymerization and the Sol-Gel Process

Stefan Spange, Hardy Müller, Christian Jäger[#] , and Cornelia Bellmann[§]*

Department of Polymer Chemistry, Institute of Chemistry, Chemnitz University of Technology, Straße der Nationen 62, 09111 Chemnitz, Germany
[#] Institute of Physics, Friedrich-Schiller-University Jena, Max-Wien-Platz 1, D-07743 Jena, Germany
[§] Institute of Polymer Research, Hohe Straße 6, Dresden, Germany

Summary: Silica particles with different morphology have been functionalized with carbon shells by different synthetic procedures. In the key step, the bare silica particles are functionalized by a specific cationic surface polymerization with furfuryl alcohol (FA). The polyfurfuryl alcohol (PFA)/silica hybrid particles have been also post-functionalized with maleic anhydride (MSA) by a Diels Alder reaction. Simultaneously occuring cationic polymerization of FA and sol-gel process with TEOS has been used for producing interpenetrating carbon-silica hybrid materials. The thermal transformation of the PFA component on silica into the carbon phase has been carried out under argon atmosphere in a temperature range from 700 °C to 900 °C. The influence of the former morphology of the silica on the homogenity of the resulting carbon layer is shown by zetapotential measurements and electron microscopic investigations.

Introduction

Carbon and silica play an important role in chemistry, physics, and materials science due to their relevance in practical applications and academic research. More than a thousand original papers, books, and reviews on this topic have been reported (see [1-3] and [4-5] and the references therein).

Carbon exhibits a very rich structural and surface chemistry. [1-3, 6-10] Various different architectures are possible because carbon can be bound in linear (sp), trigonal (sp^2), and tetrahedral (sp^3) coordination.[2] Many crystalline and amorphous forms have been synthesized such as films, fibres, glassy carbon or nanotubes. [1-3, 6-8] Graphite like materials are widely interesting because of their high electrical conductivity, intercalating capability, high thermal stability, and high resistance towards organic solvents. [6-8]

CCC 1022-1360/00/$ 17.50+.50/0

Hybrid materials consisting of a graphite like carbon phase and an inorganic oxidic component can be synthesized by different strategies. A convenient route for synthesizing the carbon component is the thermal transformation of a suitable organic precursor polymer into the carbon phase. [6-8, 11-21] Interpenetrating carbon/silica hybrid materials have been obtained by a sol-gel process of tetraalkoxysilanes using water soluble monomers or oligomers which are suitable as carbon precursor materials. [18-21] These materials are important for various applications, because each component serves as the template for the other one. Removing of any component results in a porous carbon and silica material, respectively. [20, 21]

In this paper we report on two different methods for producing carbon/silica hybrid materials. Carbon coated silica particles have been synthesized by method **A** and interpenetrating carbon/silica hybrid materials by method **B**.

A: Specific cationic surface polymerization of furfuryl alcohol (FA) on silica particles has been used for producing polyfurfuryl alcohol (PFA) coated silica hybrid precursors,[22, 23] because PFA has been well established as suitable precursor polymer for producing different kinds of carbon materials (scheme 1).

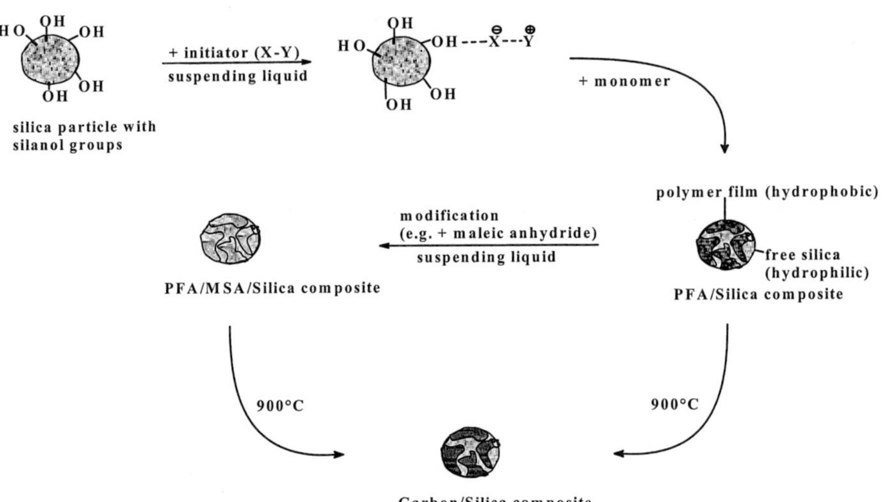

Scheme 1.   Functionalization of silica particles by cationic surface polymerization.

B: Cationic oligomerization of FA is also possible in a tetraethoxysilane (TEOS)/water mixture. Due to the high stability of the furfurylium ion, it co-exists in sol-gel processes even in the presence of water which is suitable for synthesizing oligo FA simultaneously during the sol-gel process (scheme 2).

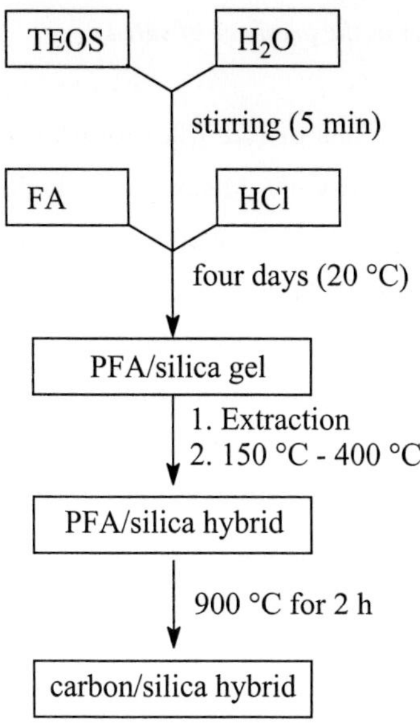

Scheme 2.   Pathway for producing carbon/silica hybrid materials from FA and TEOS.

Experimental Section

The general procedure for the synthesis of PFA modified silica particles have been reported in the refs. [23, 24].

The postderivatization of the PFA/silica particles with MSA has been carried out in xylene as solvent. The structure analysis has been previously reported in a recent paper by Günther. [25]

The carbonization of the resultant PFA/silica and PFA-MSA/silica particles has been carried out as follows. The sample was heat-treated at a linear rate of 10 °C /min to 900 °C and held there for 3h to carbonize completely the polymer on the silica surface under inert atmosphere. The carbon contents were also analyzed by quantitative elemental analysis.

Solid state NMR spectroscopy, Raman spectroscopy, scanning electron microscopy, zetapotential and BET measurements have been performed as reported in the refs. [20, 23, 24, 26].

Results and Discussion

Cationic polymerization of FA in the presence of silica

For this work we used different kind of bare silica particles for functionalization, irregular particles with random size shape [KG 60, Merck®, particle diameter: 40 - 60 µm; specific surface (BET): 423 m^2/g; high porosity; pore volume: 0.71 - 0.78 ml/g] and spherical particles [Purospher® Si 80 **(PSP)** from Merck®, particle diameter: 5 µm; specific surface (BET): 405 m^2/g], of high purity and high quality in morphology (see experimental part). The cationic polymerization of FA on the surface of silica particles as well as in the sol-gel reaction mixture and the associated structure formation of PFA are strongly dependent on the experimental conditions for the polymerization.[24] The main important influence is that of reaction temperature. A higher temperature is responsible that formation of branched and conjugated sequences are improved, whereas ether bridges formation [-(C_4H_2O)-CH_2-O-CH_2-(C_4H_2O)]- is suppressed. For this report, we have selected the results from the optimized experiments. Therefore, we have used preferably those PFA/silica hybrid batches which have been synthesized at higher reaction temperature.

The solid state MAS CP {^{1}H} ^{13}C NMR analyses of the PFA/silica hybrid particles clearly demonstrates that PFA has been formed.[27] The PFA modified silica particles can be readily chemically post-functionalized by a Diels-Alder reaction with maleic anhydride (MSA) (see experimental part). The resulting materials are denoted as PFA-MSA/silica particles. The surface structure characterization of the PFA-MSA/silica material was also carried out with solid state ^{13}C NMR spectroscopy. Figure 1 shows the assignment of the ^{13}C NMR signals found in the solid state MAS CP {^{1}H} ^{13}C NMR spectrum of PFA-Silica hybrid- and of maleic acid functionalized PFA-Silica hybrid particles (MSA-PFA/silica).

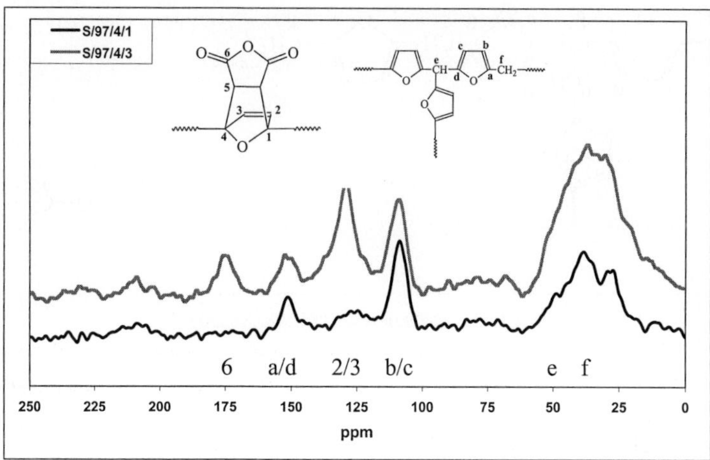

Figure 1.   Molecular structures of the PFA network on the silica particle surface and assignment of the functional groups to the signals appearing in the solid state MAS CP ^{13}C {^{1}H} NMR spectra of the hybrid particles.

The -[-C$_4$H$_2$O-CH$_2$-]- chains can be evidenced in the solid state NMR spectra by the ^{13}C signals at about $\delta = 151$ ppm, $\delta = 108$-111 ppm, and $\delta = 28$ ppm relating to the carbon atoms of the furan ring C2 / C5, C3 / C4, and the methylene bridge, respectively (see Figure1). The signal at $\delta = 39$ ppm indicates crosslinking between oligoFA sequences. [27] The structure of the resulting PFA on silica is not uniform. However, the expected -(CH$_2$-C$_4$H$_2$O)- units are mainly present beside conjugated sequences which are formed by hydride transfer and isomerization reaction during the cationic surface polymerization. The conjugated sequences are responsible for the dark brown colour of the cationically synthesized PFA/silica hybrid particles. However, the concentration of the conjugated sequences are too low to be detectable by solid state NMR spectroscopy. When FA is simultaneously co-polymerized in the sol-gel process, then the fraction of conjugated sequences formed is evidently larger. This has been suggested by the solid state ^{13}C NMR spectrum of the interpenetrating PFA/silica network (Fig. 2).

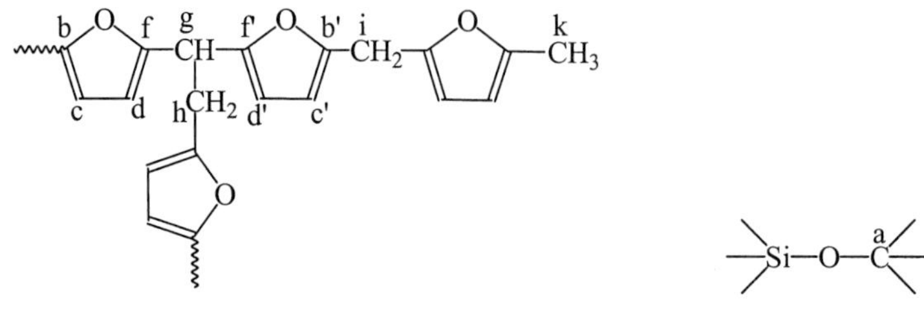

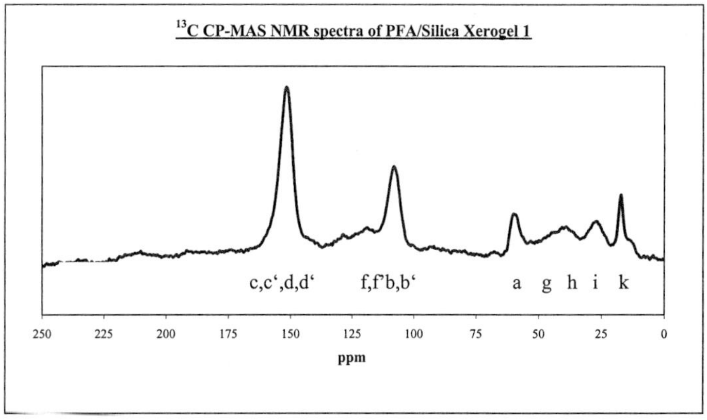

Figure 2.   Solid state MAS CP {^{1}H} ^{13}C spectrum of an interpenetrating PFA/silica hybrid material as obtained by the HCl induced simultaneously occuring cationic polymerization of FA and sol-gel process with TEOS and water at 293 K. The carbon content of the hybrid is about 18 %.

Beside the expected -[-C$_4$H$_2$O-CH$_2$-]$_n$- chain sequences, which can also be evidenced in the solid state NMR spectra by the ^{13}C signals at about δ = 151 ppm, δ = 108-111 ppm, and δ = 28 ppm, (see Figure 2), a further rather intense signal at about δ = 60 ppm indicates Si-O-C bonds [28] and the broad signal at δ = 39 ppm indicates the crosslinking between oligoFA sequences. The main part of the PFA fraction cannot be removed from the xerogel by an extraction with organic solvents (methanol, 1,2-dichloroethane) or water. Furthermore, a ^{13}C signal hinting at methyl groups is found at δ = 17.5 ppm.

The solid state ^{1}H MAS NMR spectrum of the PFA/silica hybrid xerogel gave no detailed structure information. Broad signals for the protons of the furan ring at about δ = 5.95 ppm and δ = 1.1 ppm for methyl groups are observed as well as rather wide signals at δ = 3.95 ppm which cannot be assigned to specific structure elements because of silanol groups, water, and possible TEOS residuals. The methyl groups may be formed by a hydride transfer reaction from the methylene bridges of oligoFA to cationically active oligofurfurylium intermediates. However, signals for Si-O-C bonds as well as methyl groups in the solid state NMR spectra can also be caused by TEOS residuals in the polyFA/silica xerogel. The high content of methyl head groups in the FAoligomers would be consistent with the black color of oligoFA, because the hydride transfer reaction is responsible for the formation of the conjugated sequences. The ^{13}C signals for the olefinic carbon atoms are not resoluted, but they are clearly detectable in the solid state ^{13}C $\{^1$H$\}$ NMR spectrum as a broad signal between $\delta = 120$ and $\delta = 140$ ppm. The cationically conjugated sequences are responsible for the dark green color of the PFA/hybrid gel [$\lambda = 600$ nm and a shoulder at about $\lambda = 765$ nm]. Because conjugated polymers are especially suitable precursor polymers for producing carbon, the PFA cationically produced on silica particles and among the sol-gel process, respectively, is advantaged for this purpose.

Chart 1. Theoretically possible reactions of the cationically active polyfurfurylium chain with PFA, conjugated sequences, and the silanol groups formed during the sol-gel process.

Both kind of precursor materials PFA/silica or PFA-MSA/silica and the PFA/silica xerogel have been used for the carbonization procedure in this paper.

900 °C has been found to be an optimized carbonization temperature for both kinds of precursor materials MSA-PFA/silica or PFA/silica and PFA/silica xerogels. Using this

optimized temperature, the carbonization reaction proceeds to is completion, because the IR signals derived from the precursor polymers are disappeared completely in the DRIFT-spectrum. A higher carbonization temperature than 900 °C is not suitable, because then the BET-surface area of the porous carbon/silica particles decreases significantly due to changes of the morphology of the silica frame. [24]

Carbon/silica particles

When above 12 % weight of carbon are present on the carbon/silica hybrid particles surface, the silanol groups completely disappear as seen by the lack of the single valency vibration of the silanol groups at $v = 3740$ cm^{-1} in the DRIFT spectrum.

We presume that the MSA functionalized PFA/silica particles are better suited for the carbonization reaction since the thermal cleavage of the anhydride groups should occur even at lower carbonization temperature resulting in CO_2 and aromatic rings. The surface carbonization process transforming PFA into carbon also proceeds very effectively since conjugated sequences are present in the precursor polymer. We think that the former conjugated sequences in the PFA layer on silica serve like a seed for producing carbon germs during the first stage of carbonization.

These germs are present in high concentrations which significantly improve the carbonization reaction. We think that these both facts are responsible for the high yield of the carbon phase. The formation of the carbon phase among the silicatic frame is clearly shown by Raman spectroscopy. [20, 30,31] The results of Raman as well as EPR spectroscopic investigations (the evident signal at $g = 2.00027$ [G] indicates paramagnetism) show the graphite like property of the carbon layer. The carbon/silica particles produced from MSA-PFA/silica show the highest conductivity which supports that a graphite like carbon layer has been formed. The difference of conductivity compared to carbon/silica from PFA/silica is significant. Qualitative indicative results are shown in Figure 3.

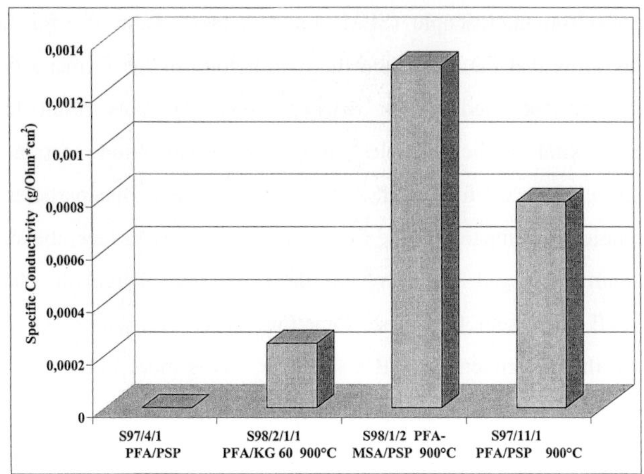

Figure 3.   Conductivity of PFA functionalized silica particles (S97/4/1) Carbon coated Purospher® particles using PFA/silica(KG 60) (S98/2/1/1), Carbon coated Purospher® particles using PFA/PSP precursor (S97/11/1) and MSA-PFA/PSP precursor (S98/1/2). (PSP = porous silica particle).

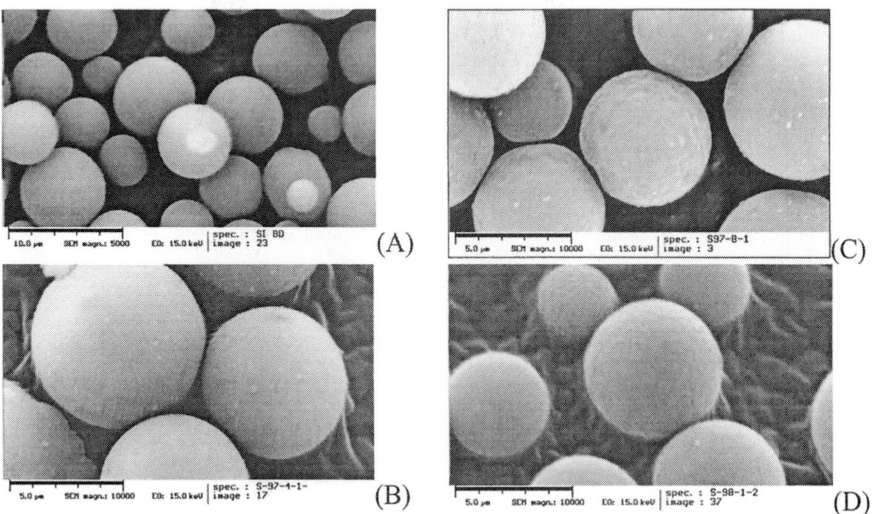

Figure 4.   Scanning electron microscopic (SEM) pictures of
(A)   bare Purospher® silica particles
(B)   PFA functionalized silica particles
(C)   carbon coated Purospher® particles using PFA/silica as precursor
(D)   carbon coated Purospher® particles using MSA-PFA/silica as precursor.

The scanning electron microscopic (SEM) pictures taken from the functionalized silica particles studies show that the shape and the morphology of the former silica frame is not changed by the surface modification reactions (Fig. 4). This result has been found independent of the kind of silica particle used for the derivatization reaction. Especially, the carbon layer derived from the former MSA functionalized PFA/silica particles show a "tennis ball" like morphology. Compared to the former silica particles surface, the size and shape of the particle remain unchanged. The IR results show, that neither free silica batches nor holes in the carbon shell are present. Of course, sometimes (very rarely) a broken hybrid particle is observed among the nonbroken ones. But we think this is more likely due to mechanically induced influences rather than caused by the modification procedure

For detecting sensitive differences to the surface properties, zetapotential measurements have been well established for characterization of bare silica and functionalized silica particles as well as carbon. [26, 29] Figure 5a shows the zetapotential plots as function of pH for differently functionalized KG 60 silica samples and Figure 5b for the Purospher® silica particles (PSP) studied in this work.

Overall, for bare silica, the IEP is observed at pH = 2.8. Functionalization of silica with PFA causes no significant shift upon the IEP. This behavior is expected and shows that the surface charges in water are mainly determined by the residual silanol groups on the PFA/silica hybrid particles surface.[29] The post-functionalization of the PFA/silica hybrid particles with MSA causes a significant shift of the IEP to lower pH, at approximately pH = 2. This effect is well in accordance with the higher acidity of maleic acid in water (pK_s = 1.92 at 298 K) compared to that of silica (pK_s = 2.65).[29]

Generally, for the carbon/silica hybrid particles a significant shift of the IEP to higher pH is observed. The shift observed is different for the two silica samples used. It takes place from pH = 2 (MSA-PFA/silica) and pH = 2.8 (PFA/silica), respectively, to pH = 4 for carbon/silica KG 60. For comparison, (non derivatized) carbon fibres show the IEP at pH = 4.5 and qualitatively similar zetapotential plots as function of pH like the carbon/silica particles. [26] This comparison shows that the macroscopic surface properties of the carbon/silica KG 60 particles are mainly determined by the carbon layer and not by the silicatic core. The small difference of the IEP measured for pure carbon (IEP at pH = 4.5)[26] compared to carbon/silica (IEP at pH = 4) is caused by accessible weakly acidic groups which are likely operative at edges and irregular sizes of the particles surface. They are not detectable by DRIFT spectroscopy.

The carbon/silica hybrid particles synthesized from PSP/silica show the IEP at pH = 4.5, which agrees with the expected value for the pure carbon phase. This result shows that a homogeneous carbon layer has been formed. Thus, the method described here is very suitable for producing carbon/silica particles with adjustable surface properties.

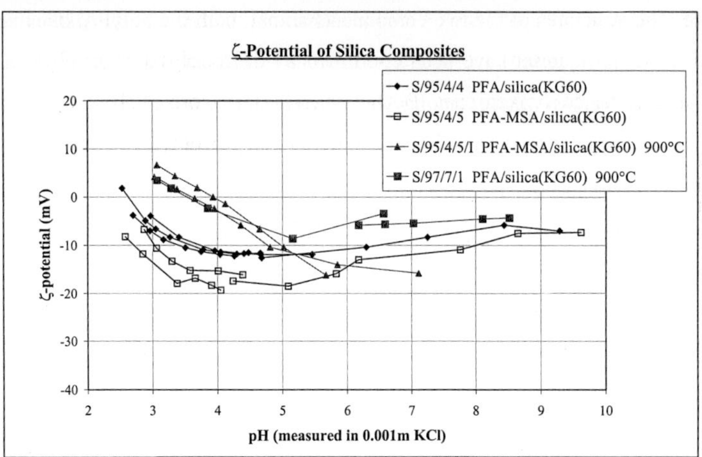

Figure 5a.   Zetapotential plots of PFA/silica(KG 60) (-◆-), PFA-MSA/silica(KG 60) (-□-) and carbon/silica(KG 60) from PFA-MSA/silica precursor (-▲-) and from PFA/silica precursor (-■-) as function of the pH of the aqueous solution.

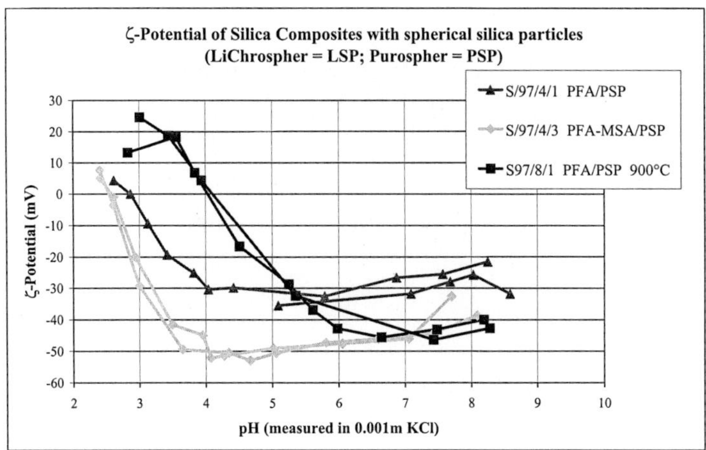

Figure 5b.   Zetapotential plots of PFA/silica(PSP) (-▲-) PFA-MSA/ silica(PSP) (-◆-) and carbon/silica(PSP) from PFA/silica precursor (-■-) as function of the pH of the aqueous solution.

Interpenetrating carbon/silica monoliths

PFA/silica xerogel hybrid materials with different compositions can be synthesized by simultaneously occurring cationic polycondensation of FA and sol-gel process with TEOS.[20, 21] Those xerogel hybrid materials are suitable precursors for producing nanostructural carbon/silica materials by their thermal induced transformation at 900 °C under inert atmosphere. The structures of the two components among both the polyFA/silica xerogel and carbon/silica hybrid material have been confirmed by different spectroscopic methods. The formation of the carbon phase among the silicatic frame is clearly shown by Raman spectroscopy. [30, 31] The two characteristic bands for the carbon phase at $\nu = 1300$ cm^{-1} and $\nu = 1550$ cm^{-1} are clearly observed. Both Raman bands have the same intensity. This result of the Raman spectra suggest that the carbon phase is graphite like and non-porous. This is confirmed by attempts of BET measurements. A measurable BET surface area could not be determined for the carbon/silica hybrid material. It is lower than 1 m^2/g. The carbon phase can be burned with oxygen. Afterwards a porous silica is obtained which has a BET surface area of about 120 m^2/g. The silica phase shows a porous structure with mesopores in the range of 20 nm to 30 nm. The remaining silica phase obtained after the burning process contains nanopores with 2-3 nm size which amounts to about 80 % of the total pore volume. Inside the pores the polymer, respectively, the carbon phase is located (white colored sections in Figure 6). As a consequence, an interpenetrating network structure is formed (see Fig. 6).

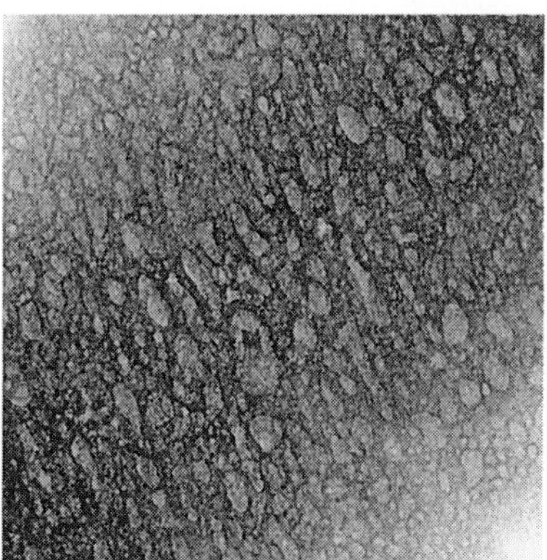

Figure 6.   Electron micrograph of an interpenetrating carbon/silica hybrid material.

Thus, during the sol-gel process the two components formed PFA and silica, respectively, serve as a template for each other, which offers the opportunity in constructing new mesoporous hybrid materials by removing the nondesired component either by HF (silica) or oxygen (carbon).

Acknowledgement

Financial support by the Fonds der Chemischen Industrie, Frankfurt/Main, and the University of Technology, Chemnitz, is gratefully acknowledged.
The authors thank Simone Kehr, Polymer Chemistry, for technical assistance and Prof. Dr. R. Holze, Electrochemistry, Technical University Chemnitz, for providing the equipment to do the conductivity measurements. We thank Dr. Manfred Friedrich, Insitute of Inorganic Chemistry, Friedrich-Schiller-University Jena, for recording and utilizing the EPR spectra.

References

(1)   J.C., Bokros, In *Chemistry Physics of Carbon*; Walker, P., L., Jr., Ed..; Marcel Dekker: New York **1969**; *Vol.5*, 1-118.

(2)   F. Diederich, Y. Rubin, *Angew. Chem.* **1992**, *104*, 1123-1292; Int. Ed. Engl. **1992**, *31*, 1101 and references cited therein.)

(3)   T. W. Ebbesen, *Acc. Chem. Res.* **1998**, *31*, 558-566.

(4)   R. P. W. Scott, *Silica Gel and Bonded Phases* , **1993**, John Wiley & Sons.

(5)   R. K. Iler, *The Chemistry of silica*, 1979 John Wiley & Sons,New York.

(6)   C.-C. Han, J.-T. Lee, R.-W. Yang, H. Chang, H., C.-H. Han, Chem. Commun. **1998**, 2087

(7)   E. D. Miller, D. C. Nesting, J. V. Badding, Chem. Mater.**1997**, *9*, 18-22.

(8)   O. Vohler, P.-L. Reiser, R. Martina, D. Overhoff, Angew. Chem. **1970**, *82*, 401-452.

(9)   E. Fitzer, *Angew. Chem.* **1980**, *92*, 375-386.

(10)   H. O. Böder, *Angew. Makromol. Chem.* **1982**, *109/110*, 125-138.

(11)   J. Rodriguez-Mirasol, T. Cordero, L. R. Radovic, J. J. Rodriguez, *Chem. Mater.* **1998**, *10*, 550.

(12)   T. Kyotani, T. Nagai, S. Inoue, A. Tomita, *Chem. Mater.* **1997**, *9*, 609.

(13)   M. Narisawa, K. Yamane, Y. Okabe, K. Okamura, *J. Mater. Res.* **1999**, *14*, 4587.

(14)   S. Han, T. Hyeon, *Chem. Commun.* **1999**, 1955.

(15)   N. Sonoba, T. Kyotani, A. Tomita, *Carbon* **1988**, *26*, 573.

(16)   T. Kyotani, N. Sonoba, A. Tomita, *Carbon* **1991**, *29*, 61.

(17)   T. Kyotani, T. Nagai, S. Inoue, A. Tomita, A., *Chem. Mater.* **1997**, *9*, 609.

124

(18)   S. M. Manocha, D. Y. Vashistha, L. M. Manocha, *J. Mater. Sci. Lett.* **1997**, *16*, 705.

(19)   M. Narisawa, K. Yamane, Y. Okabe, K. Okamura, *J. Mater. Res.* **1999**, *14*, 4587.

(20)   H. Müller, C. Jäger, P. Rehak, N. Meyer, J. Hartmann, S. Spange, *Adv. Mat.* **2000**, *12*, 1671.

(21)   D. Kawashima, T. Aihara, Y. Kobayashi, T. Kyotani, A. Tomita, *Chem. Mat.* **2000**, *12*, 3397.

(22)   A. Gandini, *Adv. Polym. Sci.* **1977**, *25*, 17.

(23)   S. Spange, *Progr. Polym. Sci.* **2000** , *25*, 781-849.

(24)   H. Müller, S. Spange, G. Marx, N. Meyer, I. Mehlhorn, C. Bellmann, Chem. Mat. 2001, submitted

(25)   H. Günther, V. Franke, Fresenius *J. Anal. Chem.* **1997**, *357 (5),* 505.

(26)   F. Simon, H. J. Jacobasch, D. Pleul, P. Uhlmann, *Progr. Coll. Polym. Sci.* **1996**, *101*, 184-188.

(27)   S. Spange, H. Schütz, R. Martinez, *Makromol. Chem.* **1993**, *194*, 153.

(28)   D. M. Hoebbel, D., M. Nacken, H. Schmidt, *J. Sol-Gel Sci. Techn.* **1998**, *12*, 169.

(29)   F. Simon, H. J. Jacobasch, S. Spange, *Coll. Polym. Sci.* **1998**, *276*, 930-939.

(30)   D. S. Knight, W. B. White, *J. Mater. Res.* **1989**, *4*, 385.

(31)   T.-H. Ko, C.-Y. Chen, *J. Appl. Polym. Sci.* **1999**, *71*, 2219.

Graft Copolymers Composed of High Molecular Weight Poly(ethylene oxide) Backbone and Poly(*N*-isopropylacrlylamide) Side Chains and Their Thermoassociating Properties

E. Hasan, K. Jankova[1]*, V. Samichkov*[2]*, Y. Ivanov*[2]*, and Ch.B. Tsvetanov**

Institute of Polymers, Bulgarian Academy of Sciences, Sofia 1113, Bulgaria
Tel.:+359-2-700-138; Fax :+359-2-707-523;
E-mail address:chtsvet@polymer.bas.bg
www.polymer.bas.bg
[1]Assen Zlatarov University, Bourgas 8010, Bulgaria
[2] Central Laboratory of Physicochemical Mechanics, Bulgarian Academy of Sciences, Sofia 1113, Bulgaria

Summary: Novel, water-soluble thermoassociative graft copolymers based on high molecular weight (HMW) poly(ethylene oxide-co-glycidol) backbone and relatively short grafts of poly-N-isopropyl acrylamide (NIPAAm) were prepared. The copolymer precursors with two architectures (block and graft) were synthesized using Ca-amide-alkoxide initiators. The OH groups in the copolymer precursors have been utilized for grafting NIPAAm using ceric ion (Ce^{4+}) redox initiation. The idea was to imprint the "smart" properties of PNIPAAm grafts into common HMW poly(ethylene oxide). The sensitive moieties undergo reversible association transitions by changing the temperature of dilute and semidilute aqueous solutions of the copolymers. Associative properties were studied by viscosity and rheology measurements. Two types of interactions, induced by heating, depending on the copolymer concentration namely intra- and intermolecular association were observed.

1. Introduction

High molecular weight poly(ethylene oxide) (HMW PEO) is widely used as thickening, flocculating, or suspending agent. One of the advantages of HMW PEO in this application is that only very low aqueous solution concentrations are required in order to achieve dramatic effects (1). Unfortunately, aqueous solutions of HMW PEO are generally very sensitive to deformation and some mechanical degradation (loss of viscosity) can occur under high shear rate. As a result an alternative route for preparing thickeners was developed using the concept of "associative polymers".

Hydrophobically associating copolymers consist of a water-soluble polymer containing a small number of hydrophobic groups. In aqueous solution, above a certain polymer concentration, intermolecular hydrophobic interactions lead to the formation of macromolecular associations. As a consequence, these materials exhibit thickening properties equivalent to those observed for higher molecular weight homopolymers. The physical links between chains are disrupted under increasing shear but they reform with decreasing shear. In this way, it is possible to avoid the irreversible mechanical degradation which occurs in very high molecular weight samples when subjected to high shear stresses. However, for all these systems, the viscosity generally decreases when temperature is raised and this can be an important drawback, especially for applications over a wide range of temperatures.

N-isopropylacrylamide has been, and continues to be, the most widely used component of reversibly thermosensitive systems in water, viz. solutions of linear PNIPAAm and swollen hydrogels of crosslinked species. The unique properties of poly(ethylene glycol)-poly(N-isopropylacrylamide) block and graft copolymers in water have recently attracted a lot of attention (2-4). Part of the interest stems from the lower critical solution temperature (LCST) of PNIPAAm in water that can be tuned close to body temperature by copolymerization and may therefore be applied in the biomedical field as a stimulus-sensitive material (5). Since the hydrophilic PEG provides steric stabilization of the block copolymer in aqueous solution, the combination of the temperature-sensitive PNIPAAm and PEG should exhibit interesting thermosensitive aggregation behavior.

In the present work we undertook the synthesis and characterization of thermoassociating water soluble graft copolymer systems based on high molecular weight PEO. The idea was to imprint some "smart" properties into common PEO by introducing responsive components consisting of relatively short segments of PNIPAAm grafts. The sensitive moieties can undergo reversible association transitions by changing the environmental conditions. This approach should lead to a very complex phase transition behavior with potential applications in numerous fields including drug delivery, chemical separations, sensors, catalysis and thickening agents.

2. Experimental

2.1.Materials

Glycidol, (G) (Aldrich) was distilled under reduced pressure prior to use. N-isopropylacrylamide, (Aldrich) was recrystallized from an hexane/acetone (10:1 v/v) and vacuum dried. Ammonium cerium (IV) nitrate was purchased from Aldrich. All other chemicals were used as received.

2.2. Synthesis of precursor copolymers

The anionic precipitation polymerization of ethylene oxide (EO) and G was initiated by calcium-amide-alkoxide (6). Calcium amide modified catalyst was synthesized directly in the reaction vessel. This involved reaction of 0.5 g Ca with 0.6 ml EO in 7.5 ml heptane and 35 ml liquid NH_3 followed by thermal treatment of the catalyst suspension at 80 °C for 1 hour.

2.2.1. Poly(ethylene oxide-co-glycidol)

The copolymerization leading to random copolymers, was carried out with the calcium catalyst at 40 °C for 5 hours. After 15 min of slow bubbling of EO through the suspension formed of the modified Ca catalyst in 85 ml heptane, a certain amount (0.5-2.5 ml) G was added dropwise for 4 hours, while the bubbling of EO continued.

2.2.2. Poly(ethylene oxide-b-glycidol)

Diblock copolymers, where the first block is HMW PEO and the second block polyglycidol, were obtained by sequential anionic polymerization of EO followed by addition of G. The highly reactive hydroxyepoxide represents a latent AB_2 monomer that can be polymerized to hyperbranched polyethers with numerous hydroxyl end groups (7).

The polymer precipitates (*2.2.1* and *2.2.2)* were isolated by filtration, washed several times with heptane and the free G was then extracted with freshly distilled diethylether. The copolymers were further purified from catalyst residue and oligomers through dialysis in distilled water for a week. A dialysis cellulose membrane (Dialysis tubing, Sigma) with a cut-off value of 12000 was used. The dialyzed solution was transferred to a flask, the water was removed by evaporation at ambient temperature and the residue was dried to constant weight in a vacuum oven at 40°C.

2.3. Synthesis of PNIPAAm graft copolymers

1 g of random or block copolymer precursor was dissolved in 100 ml of deionized water under vigorous stirring and 0.5-2 g of NIPAAm was added. The solution was transferred into a 250 ml two-neck round bottom flask equipped with condenser, gas inlet/outlet, and magnetic stirrer, heated to 30^{0}C while flushing with nitrogen for few hours. 10 ml aqueous solution of 0.11 g (0.2 mmol) cerium ammonium nitrate and 0.2 g (2 mmol) 63% nitric acid was added and the reaction was continued under stirring overnight. The reaction mixture was neutralized with an aqueous 0.25 M

NaOH. The solution was dialyzed through cellulose membrane for a week.The water was evaporated under vacuum.

3. Copolymer characterization

3.1. Spectral measurements

The copolymer structure was characterized by UV-, IR-spectroscopy and NMR. The ^{1}H and ^{13}C NMR spectra were recorded on a Bruker WM 250 instrument at room temperature, using D_2O, $CDCl_3$ or dry DMSO-d_6 as solvents. The compositions of the graft copolymers were determined by ^{1}H NMR in $CDCl_3$ from the relative intensities of the oxyethylene protons at $\delta = 3.64$ ppm and the methine protons of the isopropyl groups at $\delta = 4.00$ ppm.

3.2. Molecular mass characteristics

Molecular weights and molecular weight distributions of the samples were measured by SEC with double detection on a chromatography line consisting of a M510 pump, a U6K injector, two Ultrahydrogel columns with pore size of 120 Å and 250Å, a differential refractive index detector M 410, and a tunable absorbance detector M 486 (Waters Chromatography Division). Measurements were performed in MeOH/H_2O (15/85 v/v) solvent at 20^0C with a nominal flow rate of 0.8 ml/min and molecular weights were calculated by a "universal calibration" using PEO and polyethylene glycol narrow molecular weight standards.

3.3. Viscosity measurements

Solution viscosity was determined with an Ubbelohde viscosimeter having diameters of 0.45 mm at 25 °C and 40 °C. Intrinsic viscosity was measured with an initial concentration of 1.2 wt. %. The viscosity-average MW (M_v) of copolymers was calculated using the constants for pure PEO K=1.25×10^{-4} and α=0.78 (1). This relation has been established for PEO, and therefore it provides only a rough estimation of the molar mass of copolymers.

3.4. Rheological investigations

Solutions of 5, 7.5 and 10 % of PEO were prepared in distilled water. A 0.5-1 ml portion of each solution was placed on measuring sensor system type cone (d = 50 mm, 0.3°) and plate (d = 50 mm) of the Rheotron (Brabender). All experiments were carried out in the temperature range 20-60 °C, with an equilibrating time of 10 min for each temperature. The steady shear viscosity was obtained in the shear rate range from 3 to

100 s^{-1} and shear time 30 s. The dynamic oscillatory measurements were performed with frequency range of 0.01-5 Hz, amplitude in linear viscoelastic range of 0.01 rad. and 3 cycles for elipses evaluation.

3.5. Turbidimetry

The transmittance of 3 wt. % aqueous polymer solutions was measured by UV-VIS Specord (Carl-Zeiss-Jena, Germany) at 500 nm in the temperature range of 30 ÷ 40 ^{0}C equipped with a thermostated cell. The temperature was increased at a 0.2 ^{0}/min rate. The sample was allowed to equilibrate for at least 15 min at each temperature. Thus, the cloud point of copolymer solutions was defined.

4. Results and Discussion

4.1. Synthesis of copolymer precursors

The base-catalyzed ring-opening polymerization of glycidol was recently studied intensively by Frey and Muelhaupt (8). Intra- as well as intermolecular transfer steps subsequent to the ring-opening reaction can lead to the formation of primary alkoxide as active site, which further propagates resulting in branched structure (9). Co-polymerization of EO and G using ionic coordinative initiators by precipitation polymerization could be an easy way of getting functionalized high molecular weight PEO. Heterogeneous polymerization systems usually do not follow the pattern established in homogeneous polymerization. The two monomers have different distribution coefficients in the two phases present in the polymerization mixture. They can copolymerize in a random or blockwise fashion. This consideration imply that the percentage of the functional monomer may vary substantially in the polymer fractions formed at different time intervals. It is also possible that one of the monomers may homopolymerize with little or no incorporation of the other.

Copolymerizations of EO and G monomers were carried out in hexane using anionic coordination mechanism. In all cases low MW products with predominant content of G occurred during the reaction: NMR spectra of the copolymers before and after dialysis differ significantly in the G content. The mole fraction of G incorporated in the copolymer was calculated from the phenyl group content after reaction of the polymer with phenyl isocyanate (PhNCO). The result presented in Table 1 clearly show that only a small fraction of G was incorporated in the HMW copolymer. Thus it was possible to compare the intrinsic viscosity of PEO derivative with those of pure PEO.

Table 1. Synthesis and properties of copolymer precursors.
Catalyst: 0.5g Ca as amide-alkoxide; reaction temperature 40 °C; hexane 80 ml; reaction time 5hrs.

№	Type of copolymer	Glycidol ml	Yield g	$[\eta]^{30}_{H20}$	Mv $.10^6$	-[EO]-/-[G]-*
P1	Poly(EO-co-G) statistical	2	22	7.3	1.3	25
P2		2	12**	5.8	0.9	150
P3		1.5	22	6.4	1.1	20
P4	***Poly(EO-b-G)	1.5	50	7.6	1.4	50

* The ratio was determined by reaction of polymer with PhNCO, followed by twofold polymer precipitation in ether and UV analysis described in detail in (10)
** Reaction time 2 hours
*** Sequential polymerization: 4 hours blowing of EO followed by addition of glycidol in small portions for another 2 hours (10)

4.2. Grafting of NIPAAm onto copolymer precursors

The use of the ceric-cerous redox system to initiate graft copolymerization of vinyl monomers with cellulose has been reported by a number of investigators (11, 12). The mechanism of the copolymerization reactions has been reported to be a free-radical process in which the transfer of electrons from the hydroxyl groups of the cellulose to the ceric ion results in the formation of a free radical on the cellulose molecule (13). Due to its ease of application as well as its grafting efficiency, the Ce^{IV} ion method has gained considerable importance in the grafting reaction with other polymers bearing hydroxyl groups.

The OH groups in the EO/G copolymer precursors have been utilized for grafting NIPAAm using ceric ion (Ce^{4+}) redox initiation. A series of water-soluble graft copolymers composed of poly(EO-co-G) or poly(EO-b-G) backbones and PNIPAAm grafts were obtained under the conditions described in the Experimental Section.(Scheme 1 and Scheme 2)

Scheme 1.Poly(EO-co-G-g-NIPAAm)

Scheme 2.Poly(EO-b-G-g-NIPAAm)

It is useless to compare the molecular-weight characteristics of the graft copolymers with those of the starting material, since employing nitric acid in the graft copolymerization causes some cleavage of the polyether macromolecules. As a result of this side reaction the molecular weights of grafted with NIPAAm copolymers changed with one order of magnitude as shown in Table 2.

Table 2. Synthesis of grafted with NIPAAm copolymer: 1g copolymer precursor dissolved in 100 ml distilled H_2O; reaction time 16 hrs; reaction temperature 30 °C; $(NH_4)_2Ce(NO_3)_6$ - 0.2 mmol.

№	Copolymer	NIPAAm	Yield	$[\eta]_{H2O}^{25oC}$	Mn	MWD	-[EO]-/-[NIPAAm]-	
	precursor*	g	g		$.10^5$		IR	^{1}H NMR
	Poly(EO-co-G)							
P5**	P1	1	1.1	0.93	1.1	2.8	63	57
P6**	P2	2	1.15	0.87	1	2.5	26	21
P7	P2	1	1.2	2.00	1	2.3	5.3	4,6
P8	P3	0.5	1.12	1.80	0.9	2.2	4.8	4.2
P9	P3	2	1.4	2.13	1.2	2.4	4.8	4.2
	Poly(EO-b-G)							
P10	P4	0.5	1.08	3.75	1.7	3.1	13	10
P11	P4	0.8	1.15	4.10	1.8	3.2	16	12
P12	P4	1.5	1.37	3.90	2.1	3.4	14	10

* Samples of precursors, see Table 1.

** Reaction time 8 hrs; reaction temperature 60 °C.

Figure 1 shows the ^{1}H NMR spectrum of the graft copolymer in D_2O. The phase transition of the graft copolymers is associated with the decrease of the PNIPAAm chain mobility. This process is clearly shown in Fig. 1.

The chemical shift assignments for all the carbons are summarised in Table 3.

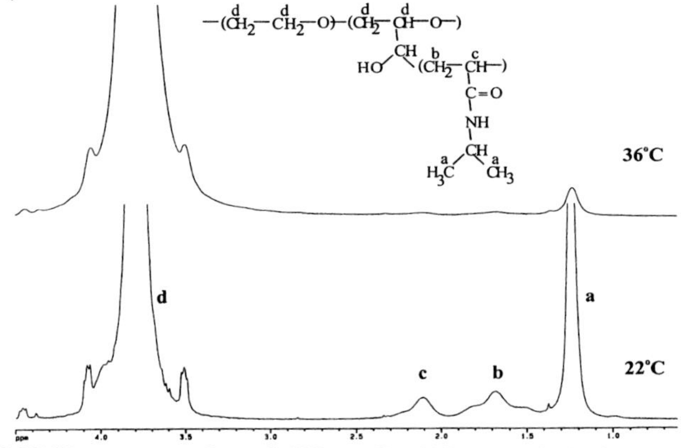

Fig. 1. ^{1}H NMR spectrum of   poly(EO-co-G-g-NIPAAm) (P8) in D_2O. Effect of temperature on the intensities of the oxyethylene backbone protons and the protons of the PNIPAAm grafts.

It is extremely difficult to consider the mechanism of grafting, since different OH groups are present in the starting material (see Scheme 1 and 2), which additionally causes broadening of the graft copolymer polydispersity.

Table 3. Chemical shift assignments for ^{13}C peaks in the graft copolymer P11

Type of carbon	Chemical shift (ppm)
- CH_3 (PNIPAm)	22.79
- CH_2 (PNIPAm)	34.68
- CH (PNIPAm)	42.97
- CH_2 (PEO)	70.79
- C=O (PNIPAm)	176.6

4.3. Phase transition

The transmission curves for the statistical and block graft copolymers are compared in Figure 2. When the temperature reaches the cloud point temperature (T_{cp}), an abrupt increase in turbidity is observed for all samples.

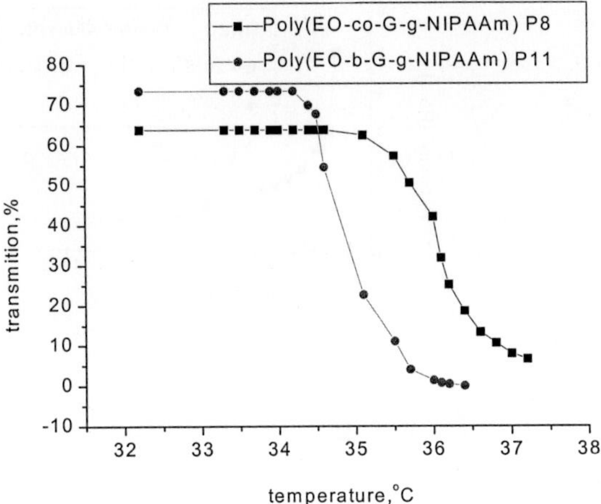

Fig. 2. Cloud point of a 3 % aqueous solution of EO/G copolymers with different architecture, grafted with NIPAAm.

The origin of this turbidity behavior arises because PNIPAAm sequences can self-assemble in aqueous solutions in micellar-like structures, stabilized by the HMW PEO blocks. It is worth noting that all copolymers do not form two well-separated phases even at temperatures much higher than T_{cp}. Instead, they appear in the form of turbid suspensions, stable for several hours. The microscopic phase separation, detected by the appearance of turbidity, is shifted to higher temperatures compared to the LCST of the PNIPAAm homopolymer (33°C). The onset of transmission curves depends on the copolymer architecture. Thus, the cloud point temperature of the random copolymer is higher than that of the blockcopolymer.

4.4. Rheological properties

4.4.1. Semi-dilute solution properties

The viscosities of semidilute graft copolymer solutions were found to decrease with shear rate which is typical for non-Newtonian systems. This kind of shear-thinning flow behavior is well known from aqueous associating polymers (14). Figure 3 shows the viscosity as a function of shear rate of a 5 wt.% solutions of poly(EO-co-G-g-NIPAAm) and poly(EO-b-G-g-NIPAAm). The viscosity of the studied systems displays a remarkable decrease up to $\gamma \approx 10$ s^{-1} and at higher shear rates the changes are not significant.

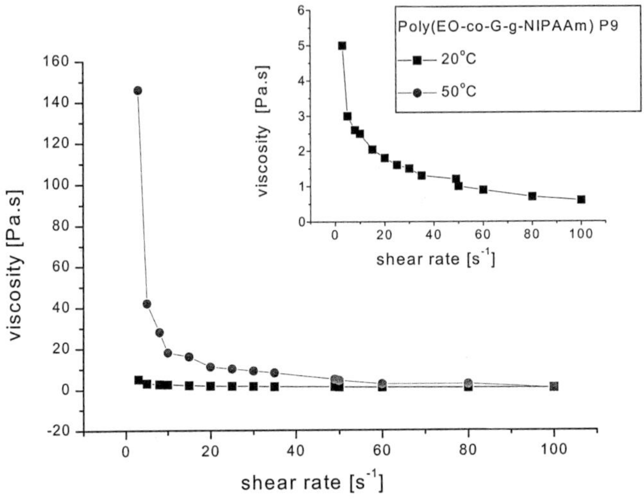

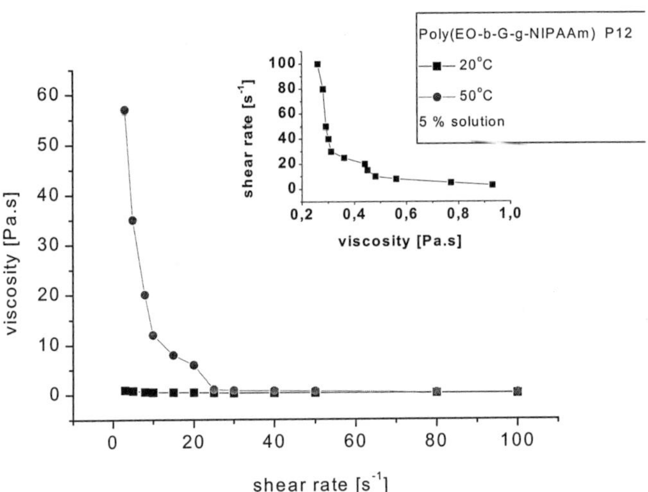

Fig. 3. Variation of the viscosity with shear rate for 5 wt.% solutions of poly(EO-co-G-g-NIPAAm) and poly(EO-b-G-g-NIIPAAm).

Figure 4 shows the reduced viscosity versus temperature plots for aqueous solutions of both copolymer structures at a shear rate of $3s^{-1}$. The viscosity increases at temperatures higher than 40 °C. Typical thermothickening behavior was observed with maximum at 50 °C. The magnitude of the thermothickening effect obtained with poly(EO-co-G-g-NIPAAm) is much greater then that observed with poly(EO-b-G-g-NIPAAm). The magnitude of thickening is almost 3 decade of viscosity for poly(EO-co-G-g-NIPAAm), and 2 decade of viscosity for poly(EO-b-G-g-NIPAAM). The onset of the thermo-thickening behavior, or the transition temperature (T_{ass}) which corresponds to the self-assembling of PNIPAAm grafts, does not fully agree with the value issued from the turbidity measurements. When the temperature reaches T_{ass}, the PNIPAAm side chains start to self-aggregate while the PEO backbone stabilize the PNIPAAm phase at the microscopic level. In semi-dilute solutions, the PNIPAAm microdomains are connected through the highly soluble PEO chains into a physical 3-dimensional network with the expected viscoelastic properties. The macroscopic transition, i.e. viscosity, proceeds continuously with increasing temperature. At T > 50°C, the viscosity decreases upon heating obviously due to inverse solubility-relationship of PEO chains. We have verified the reversibility of the thermothickening process by comparing the two plots obtained on heating and cooling. The two curves are very similar and we can conclude that in our experimental conditions, each step of the association process results in a true thermodynamic equilibrium.

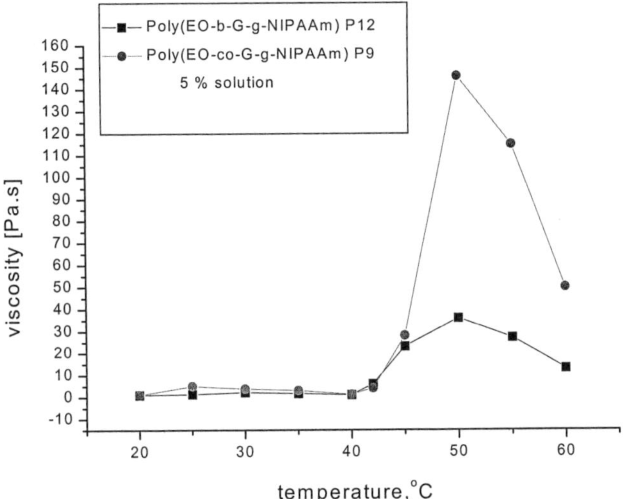

Fig. 4. Typical variation of steady flow viscosity of 5 % aqueous solutions of poly(EO-b-G-g-NIIPAAm) (P12) and poly(EO-co-G-g-NIPAAm) (P9); shear rate 3 s^{-1}.

The evaluation of the dynamic moduli provides important information about the balance between viscous and elastic forces as well about the structural changes after long relaxation periods (15). Dynamic oscillatory experiments show that at higher concentrations (7.5 wt.%) of aqueous solutions the studied systems demonstrate viscoelastic behavior. Figure 5 shows the typical frequency dependence of the dynamic moduli for the aqueous solutions of poly(EO-co-G-g-PNIPAAm).

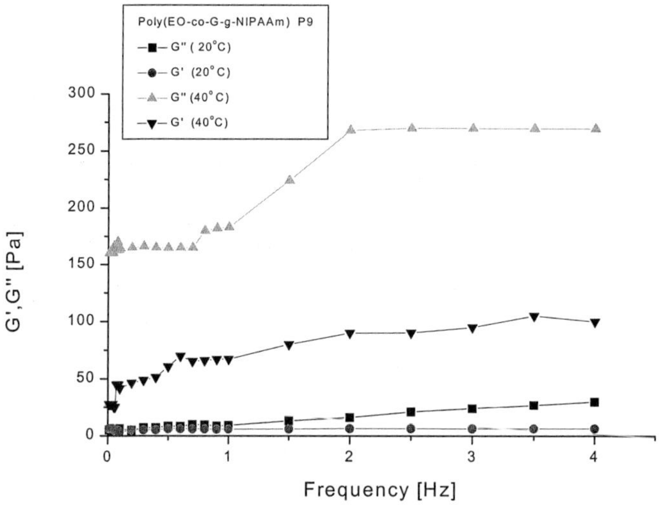

Fig. 5. Frequency dependence of the dynamic moduli of 7.5 % aqueous poly(EO-co-G-g-NIPAAm).

At 20 °C the loss modulus G'' increases with frequency when the storage modulus G' reaches a plateau after ω = 0.5 Hz. At 40 °C G' and G'' reach plateau after ω = 2.0 Hz with significant increase of their value. These results indicate that the interchain association of the copolymers is favored by temperatures higher than 40 °C.

4.4.2. Dilute solution properties

The drastic viscosity decrease in the dilute aqueous polymer solutions at 40 °C and beyond 0.5-0.7 wt. % (concentrations below the critical for coil overlapping) can be ascribed to the intrachain self-assembled association.The viscosity results plotted in Figure 6 can be compared to the monomolecular micellization where the hydrophobic PNIPAAm grafts are clustered towards the center of the shrunk coil and the hydrophilic

PEO backbone is the corona-forming moiety. Thus the polymer coils adopt a rather compact conformation, due to the formation of intrachain aggregates.

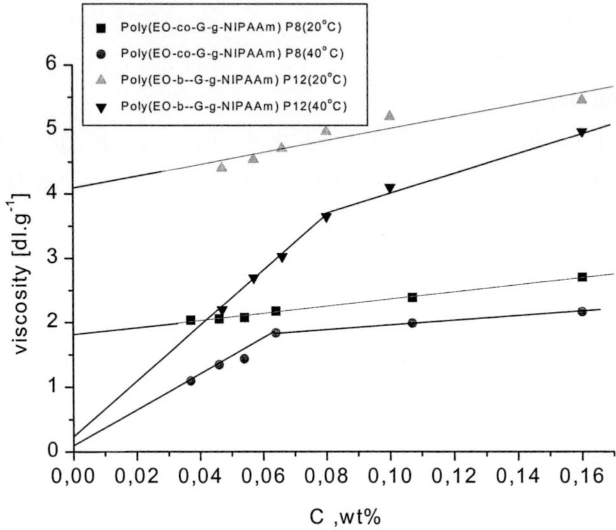

Fig. 6. Intrinsic viscosity [η] as a function of copolymer concentration at temperatures below and above the LCST of PNIPAAm grafts.

6. Conclusions

In this paper we reported the synthesis and solution properties of novel thermoassociative water-soluble graft copolymers comprising high molecular weight poly(ethylene oxide) grafted with poly(N-isopropylacrylamide) side chains.

The synthesis of thermoassociative graft copolymers has been described through a two-step procedure: 1) the synthesis of copolymer precursors of functional HMW PEO carrying OH-groups with two architectures : poly(EO-co-G-g-NIPAAm) and poly(EO-b-G-g-NIPAAm); 2) the grafting of NIPAAm onto these precursors using Ce^{IV} initiators.

From the set of PNIPAAm grafted PEO copolymers in aqueous solution we can separate two types of interactions, induced by heating, depending on the copolymer concentration that is to say intra- and intermolecular association.

In semidilute conditions, aqueous solutions of PEO copolymers start to exhibit associative properties at temperatures slightly higher than the LCST of PNIPAAm grafts. In this aggregation regime, PNIPAAm side chains begin to self-assemble forming loose clusters which behave as physical crosslinks. From the viscoelastic point of view both the number of elastically active chains of the physical network and the life time of the stickers increase.

Acknowledgements

Financial support from the National Fund "Scientific Research" (project X-808) is gratefully acknowledged. The authors thank to Prof. A.H.E. Mueller, University of Bayreuth, for helpful discussions.

1. Bailey, Jr. F.E., and Koleske J.V., *Poly(ethylene Oxide)*, Academic Press, New York, 1976, pp. 97-102
2. Yoshioka, H., Mikami, M., Mori, Y., Tsuchida, E., J. Macromol. Sci., Pure Appl.Chem., 1994, **A31** (1), 109
3. Topp, M.D., Dijkstra, P.J., Feijen, J., Macromolecules , **30**, 8518 (1997)
4. Qiu, X., Wu, C., Macromolecules , **30**, 7921 (1997)
5. Huglin M.B., Liu Y., Velada, J.L., Polymer, **38**, 5787 (1997)
6. Tsvetanov Ch.B., Dimitrov I., Doytcheva M., Petrova E., Dotcheva D., and. Stamenova R., *Application of Anionic Polymerization Research,* Roderic P. Quirk, Ed., ACS Symposium Series **696** (1998), ACS, Washington, DC, p. 236
7. Sunder A., Frey H., and Muelhaupt R., Macromol. Symp., **153**, 187 (2000)
8. Sunder A., Hanselmann R., Frey H., and Muelhaupt R., Macromolecules, **32**, 4240 (1999)
9. Kizerow D., Prohazka K., Ramireddy C., Tuzar Z., Munk P., and Webber S.F., Macromolecules, **25**, 461 (1992)
10. Dimitrov Ph. Hasan E., Rangelov S., Mueller A.H.E., and Tsvetanov Ch. B., in preparation
11. Iwakura Y. Kurosaki T., and Imai Y., J. Polymer Sci., **A3**, 1185 (1965)
12. Mino G. Kaizerman S., and Rasmussen E., J.Am.Chem.Soc., **81**, 1404 (1959)
13. Immergut E.H., in *Encyclopedia of Polymer Science and Technology*, Vol.3, H.F. Mark, N.G. Gaylord, and N.M. Bikales, Eds., Interscience, NY, 1965, p. 242
14. Hill A., Candau F., and Selb J., Macromolecules, **26**, 4521 (1993)
16. Whorlow R.W., *Rheological Techniques,* Second Edition, Ellis Horwood Limited 1998, Chichester . Engl.

Morphological Changes in ABC Triblock Copolymers by Chemical Modification

Volker Abetz[*], Kirsten Markgraf, Valéry Rebizant

Makromolekulare Chemie II, Universität Bayreuth, 95440 Bayreuth, Germany

Summary: Various SBT triblock copolymers with S being polystyrene, B being polybutadiene, and T being poly(*tert*-butyl methacrylate) and their saponified analogues, SBA with A being poly(methacrylic acid) are characterized in terms of their morphology. The chemical modification of the third block leads to a change of the overall morphology observed in solution cast films, which is interpreted as a consequence of the change of the incompatibility between the different components and the solvent.

Introduction

Block copolymers are well-known for their self-assembly into crystal-like lattices on a mesoscopic length scale. The large number of independent system variables like volume fractions and segmental interaction parameters between the different components leads to a richer morphology scheme in ternary ABC triblock copolymers as compared to linear binary block copolymers.[1,2] Quite some work has been published with the aim to understand the morphological behaviour of ABC triblock copolymers as a function of the relative composition of the different components[1,3,4,5] and also for various block sequences.[6,7] While some examples for morphological changes as a result of chemical modification of one block were given,[1,8,9] we are not aware of systematic studies on this. In the case of simultaneous modification of two blocks it was shown for example that the hydrogenation of IBS triblock copolymers (with I and B being polyisoprene and poly(1,2-butadiene), respectively) leads to EPEBS triblock copolymers (with EP and EB being poly(ethylene-alt-propylene) and poly(ethylene-co-butylene), respectively) showing other morphologies within the same morphological scheme, i.e. the increasing incompatibility between the two elastomeric blocks leads to a reduction of curvature between them and polystyrene.[10] Depending on the relative block lengths of S, the triblock copolymers exhibit a disordered state, spheres or cylinders in an IB matrix and after hydrogenation spheres, cylinders or a double gyroid of S in an EPEB matrix are formed.

The introduction of a few transition metal complexes into the middle block of a lamellar SBM triblock copolymer (with M being poly(methyl methacrylate)) with a short middle block leads to the formation of double gyroid or cylindrical morphologies.[9]

CCC 1022-1360/00/$ 17.50+.50/0

Whereas in these systems thermodynamics seems to control the morphological changes, there are also observations on the influence of solvent effects on the film casting process of block copolymers.[11,12] Upon those the most spectacular one was the so called knitting pattern morphology, which resulted from hydrogenation of asymmetric lamellar SBM triblock copolymers.[8] In this case most likely that the significant decrease of the solubility of the middle block is responsible for the morphological change. While this morphology was found in samples cast from chloroform, the same block copolymer cast from toluene solution shows lamellar morphologies. Also, an SBM triblock copolymer with a B matrix formed different morphologies depending on the solvent.[13] For this system a new hexagonal arrangement of microphase separated S and M cylinders in a B matrix was discussed.

In this contribution we will consider ABC triblock copolymers, where only one end block is chemically modified. These are polystyrene-block-polybutadiene-block-poly(*tert*-butyl methacrylate) SBT triblock copolymers, which are converted into the corresponding SBA triblock copolymers with A being poly(methacrylic acid) via hydrolysis of the poly(*tert*-butyl methacrylate) endblock.

Experimental

The SBT triblock copolymers have been obtained using living sequential anionic polymerisation. The details of the synthesis have been described elsewhere.[14,15] Polymer analogous saponification of the T block was achieved by dissolution of the triblock copolymer in a mixture of dioxane and aqueous HCl.[16,17] The solution was kept under reflux conditions for 8 hours and then the polymer was precipitated, washed and isolated. The nomenclature of the block copolymers ($A_xB_yC_z^M$) is as follows: subscripts denote the weight fractions of the corresponding block and the superscript gives the number averaged molecular weight in kg/mol. Polydispersities were obtained using size exclusion chromatography calibrated with polystyrene standards and was in the range of $M_w/M_n < 1.1$.

For morphological characterisation films of the SBT triblock copolymers were cast from solution in tetrahydrofuran (THF) or chloroform (CHCl$_3$), respectively, and the SBA triblock copolymers were cast from solutions in THF over several weeks. The samples were then further dried in vacuum at 140 °C for several days. Transmission electron micrographs were obtained from a Zeiss 902 transmission electron microscope operating at 80 kV in the bright field mode. Ultrathin sections of the samples were obtained using a Reichert ultramicrotome equipped with a diamond knife. Vaporous OsO$_4$ and RuO$_4$, respectively, were used for staining the samples. In the transmission electron micrographs the B domains appear dark, the S domains appear grey and the T and A domains remain unstained.

Results and Discussion

First, the TEM micrographs of SBT and the corresponding SBA triblock copolymers having more or less similar volume fractions of the two outer blocks ($\phi_S \approx \phi_{T,A}$) are presented (Fig. 1). For increasing volume fractions of the outer blocks the morphology of the SBT triblock copolymers ranges from spheres, tetragonally packed cylinders, a cocontinuous gyroid to lamellae. The morphological sequence observed for these SBT triblock copolymers is similar to the one found for polyisoprene-block-polystyrene-block-poly(2-vinylpyridine) triblock copolymers studied by Mogi et al.[4] The two end blocks are strongly incompatible with each other and therefore do not form common mixed microdomains. The results can be understood in the following way: The different blocks microphase separate from each other when still in solution and the solubility of both endblocks are comparable, i.e. both swollen endblocks occupy volumes of comparable size. So spheres (cylinders, gyroids, lamellae) of both endblocks appear at the same time during the evaporation process and thus occupy positions with the same symmetry properties in the bulk morphology.

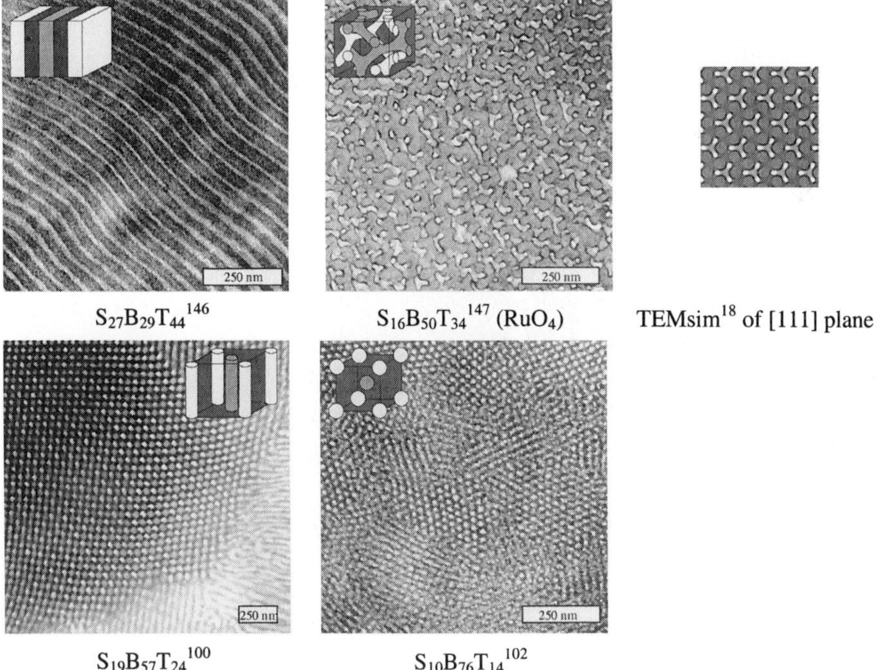

$S_{27}B_{29}T_{44}$[146]

$S_{16}B_{50}T_{34}$[147] (RuO$_4$)

TEMsim[18] of [111] plane

$S_{19}B_{57}T_{24}$[100]

$S_{10}B_{76}T_{14}$[102]

Fig. 1 Morphologies of SBT triblock copolymers

142

The SBA triblock copolymers were also cast from THF (Fig. 2). However, THF is a fairly bad solvent for poly(methacrylic acid) (e.g. water would be a good solvent, but is a nonsolvent for the other two blocks). When the A block is very short (as it is in $S_{38}B_{56}A_6{}^{66}$), it will be dissolved in a reasonably good way, so the morphology of the block copolymer will form at very high concentrations, where both S and A will probably separate in a highly entangled swollen state into similar microdomains of spherical shape. For SBA triblock copolymers with larger A block it will precipitate from the solution first, while S and B are still dissolved. In this case A will form cylindrical domains on a hexagonal lattice, while the swollen S and B blocks form the corona. At larger concentrations S will microphase separate from B and locate in cylindrical domains in the corners of the Wigner Seitz cell of the first hexagonal lattice defined by the A cylinders (Fig. 3). The same hexagonal motif was discussed by Brinkmann et al. for an SBM triblock copolymer.[13] At higher amounts of A all three blocks self-assemble into lamellar domains, as does the corresponding SBT triblock copolymer.

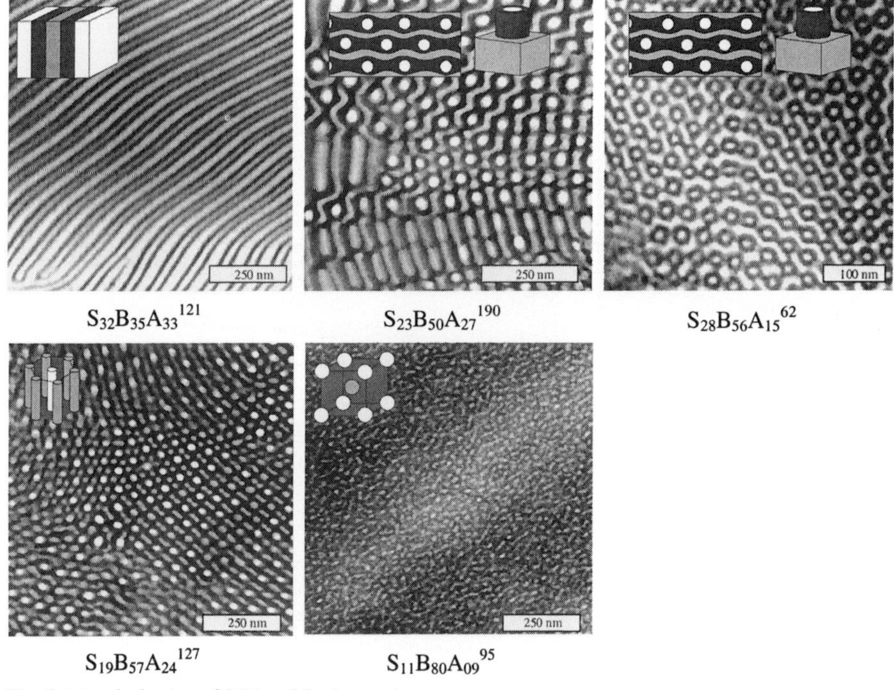

$S_{32}B_{35}A_{33}{}^{121}$ $S_{23}B_{50}A_{27}{}^{190}$ $S_{28}B_{56}A_{15}{}^{62}$

$S_{19}B_{57}A_{24}{}^{127}$ $S_{11}B_{80}A_{09}{}^{95}$

Fig. 2 Morphologies of SBA triblock copolymers

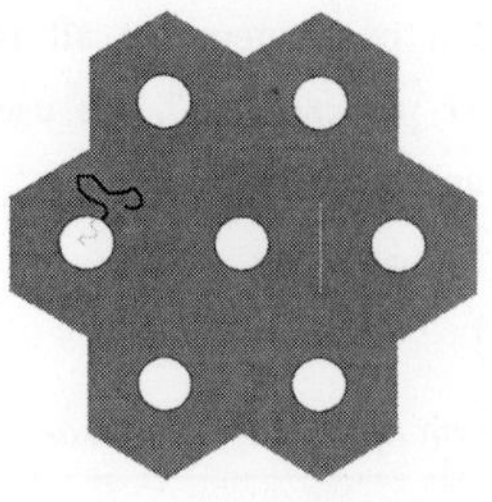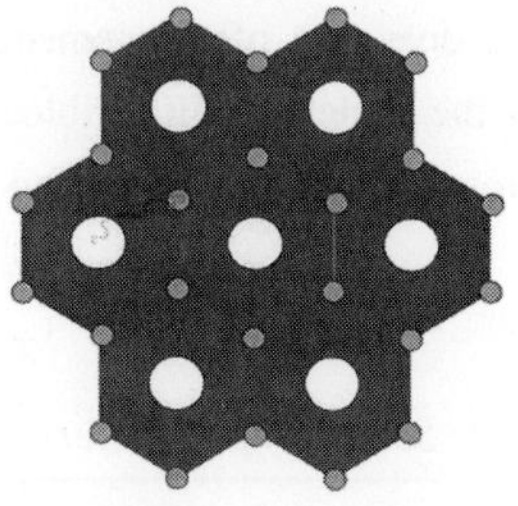

Fig. 3 Development of the hexagonal morphology by subsequent microphase separation of A (white circles) and S domains (grey circles) from the solution

For an SBT triblock copolymer with large amounts of S and B and a very short T block a lamellar structure is observed in a film cast from $CHCl_3$, where T spheres are embedded within the B lamellae (Fig. 4 left). The corresponding SBA triblock copolymer was cast from THF and forms a gyroid morphology (Fig. 4 middle). Here, the morphological change can also be attributed to the solubility of the different components. The T domain separates together with B and S and due to the fairly large amount of the latter a lamellar morphology is established with T forming small domains within the B lamellae. For the corresponding SBA triblock copolymer, the A domain microphase separates first and could possibly form either a cubic lattice of spheres (if B and S are still completely miscible at that stage) or will exhibit a gyroid network (if B and S are already weakly segregated from each other at the point, when A segregates from the solution). In the latter case S forms the other gyroid and B fills the space between the two gyroids. This interpretation is supported by TEM simulation (Fig. 4 right).

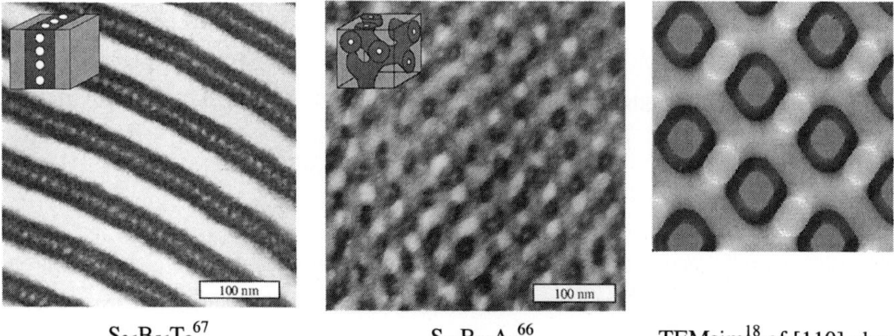

$S_{36}B_{56}T_8{}^{67}$ $S_{38}B_{56}A_6{}^{66}$ TEMsim[18] of [110] plane

Fig. 4 Morphologies of an SBT and an SBA triblock copolymer with a short third block

Since literature does not offer segmental interaction parameters for all types of binary interactions in the system "ABC triblock copolymer plus solvent", we use the solubility parameter approach following Hildebrand as a crude approximation: [19]

$$\chi \propto \left(\delta_i - \delta_j\right)^2$$

Table 1: Solubility parameters δ for estimation of interaction parameters χ

Polymer	Solubility parameter at room temperature[19,20] δ [MPa]$^{1/2}$
Polystyrene (S)	18.5
Poly(1,2-butadiene) (B)	17.4
Poly(*tert*-butyl methacrylate) (T)	18.0
Poly(methacrylic acid) (A)	21.9
Chloroform (CHCl$_3$)	19.0
Tetrahydrofuran (THF)	18.6

Comparing the solubility parameters in Table 1 it becomes obvious that the strongest incompatibility occurs between A and the other components. Thus it is likely that A microphase separates first from solution, while S and B are still soluble in it. The interactions between S, B, and T on one side and THF or CHCl$_3$ on the other are slightly different from each other and we can consider these two solvents to be much less selective for that system.

Thus there is strong evidence that solubility effects are responsible for the observation of the partly different morphological behaviour of solution cast films of SBT and SBA triblock copolymers.

Conclusions

Our results on solution cast SBT and SBA triblock copolymers demonstrate clearly the strong influence of the solvent on the final morphology. The chemical modification from T to A mainly influences the degree of selectivity of the solvent, while the increase of incompatibility towards the other two blocks (S, B) seems to be of less importance. This is probably due to the rather strong segregation of the different blocks in both SBT and SBA triblock copolymers. On the other hand, the rather different degree of swelling as a result of the selectivity of the solvent especially in the SBA system plays the key role in morphology determination.

Acknowledgement

V.A. and K.M. are very grateful to R. Stadler for inspiring discussions at the beginning of this work. The authors thank A. Göpfert for performing the TEM measurements and acknowledge fruitful discussions with I. Erukhimovich (Moscow State University), F. Court (Atofina) and H. Schmitz. This work was supported by the Deutsche Forschungsgemeinschaft (DFG) through SFB 481, the Bayreuther Institut für Makromolekülforschung (BIMF), Atofina, German Israeli Foundation (GIF) and INTAS-RFBR 99-01852.

References

1 R. Stadler, C. Auschra, J. Beckmann, U. Krappe, I. Voigt-Martin, L. Leibler, *Macromolecules* **28**, 3080 (1995)

2 F. S. Bates, G. H. Fredrickson, *Physics Today* **52**, 32 (1999)

3 G. Riess, G. Hurtrez, P. Bahadur, *Block copolymers*, in *Encycl. Polym. Sci. Eng.* **2**, 324 (1985)

4 Y. Mogi, M. Nomura, H. Kotsuji, K. Ohnishi, Y. Matsushita, I. Noda, *Macromolecules* **27**, 6755 (1994)

5 U. Breiner, U. Krappe, V. Abetz, R. Stadler, *Macromol. Chem. Phys.* **198**, 1051 (1997)

6 V. Abetz, R. Stadler, *Macromol. Symp.* **113**, 19 (1997)

7 H. Hückstädt, A. Göpfert, V. Abetz, *Polymer* **41**, 9089 (2000)

8 U. Breiner, U. Krappe, R. Stadler, *Macromol. Rapid Commun.* **17**, 567 (1996)

9 L. Bronstein, M. Seregina, P. Valetsky, U. Breiner, V. Abetz, R. Stadler, *Polym. Bull.* **39**, 361 (1997)

10 C. Neumann, D.R. Loveday, V. Abetz, R. Stadler, *Macromolecules* **31**, 2493 (1998)

11 Y. Matsushita, K. Yamada, T. Hattori, T. Fujimoto, Y. Sawada, M. Nasagawa, C. Matsui, *Macromolecules* **16**, 10 (1983)

12 Y. Funaki, K. Kumano, T. Nakao, H. Jinnai, H. Yoshida, K. Kimishima, K. Tsutsumi, Y. Hirokawa, T. Hashimoto, *Polymer* **40**, 7147 (1999)

13 S. Brinkmann, R. Stadler, E. L. Thomas, *Macromolecules* **31**, 6566 (1998)

14 C. Auschra, R. Stadler, *Polym. Bull.* **30**, 257 (1993)

15 T. Goldacker, V. Abetz, R. Stadler, I. Erukhimovich, L. Leibler, *Nature* **398**, 137 (1999)

16 C. Ramireddy, Z. Tuzar, K. Prochazka, S.E. Webber, P. Munk, Macromolecules **25**, *2541* (1992)

17 R. Bieringer, V. Abetz, A.H.E. Müller, *European Physical Journal E : Softmatter (EPJ E)* **5**, 5 (2001)

18 TEMsim 1.6 from J. Hofmann (MSRI, Berkeley)

19 A. F. Barton, "CRC Handbook of Polymer Liquid Interaction Parameters and Solubility Parameters", CRC Press, Boca Raton 1990

20 J. Brandrup, E. H. Immergut, *"Polymer Handbook"*, 3rd ed., Wiley, New York 1989

Labile Hyperbranched Polymers Used as Nanopore-Forming Agents in Polymeric Dielectrica

B. Voit[1], M. Eigner[1], K. Estel[2], C. Wenzel[2], J. W. Bartha[2]*

[1] Institut für Polymerforschung Dresden e.V., Hohe Strasse 6, D-01069 Dresden, Germany

[2] Technische Universität Dresden, Institut für Halbleiter und Mikrosystemtechnik, D-01062 Dresden, Germany

Summary: Hyperbranched polyesters containing labile triazene units in the main chain have been synthesized and characterized. These structure decompose photochemically upon UV irradiation and thermally above 160 °C. Homogeneous incorporation of these globular, functional polymers in thermostabile polymer matrices (polyimides and BCB polymers) was possible. Smooth films from blends of thermolabile matrix and up to 50% hyperbranched polymers could be prepared and have been studied as insulating material in microelectronic multilayer applications.

Introduction

Special hyperbranched, arborescent graft, or hyperbranched-co-star polymers which e.g. degrade thermally at a certain temperature can be used as pore forming templates in a temperature stable matrix.[1] This can lead to porous nanomaterials which is e.g. of interest in chromatography and for the formation of aerogels and xerogels. Hence, Boury et al.[2] described the use of dendrimers and arborols with carbosilane cores for the preparation of hybrid xerogels. In this regard, Muzafarov et al. [3] studied the degradation behavior of hyperbranched poly(bis(undecenyloxy)methylsilane)s. Of high interest is also the preparation of nanoporous polymers with low dielectric constants for use as novel interlayer dielectric material (ILD)[1,4,5]. The high demand for faster computers, higher storage capacities and at the same time smaller devices requires the development of new dielectric materials with dielectric constants below 2.5 or even 2 for chip production of very high structure density. Already realized porous systems have reached dielectric constants from $\varepsilon_r = 2.5$ [6] to approximately 1 as found for aerogels [7]. The decrease of ε_r is given by the inclusion of air filled cavities which have been formed by degradable star or dendritic polymers or by phase separated block

  CCC 1022-1360/00/$ 17.50+.50/0

copolymers with labile segments[9,10]. A major problem of nanofoams is their processability or compatibility with the integrated circuits (mechanical stability, adhesion, stability towards CMP process or selective etching). By the use of labile polymers as pore forming material in a stable polymer matrix, one should be able to introduce porosity into ILD polymers already being used for microelectronics. Important requirements for the success of the concept is a high compatibility of the hyperbranched polymers - or any globular shaped polymer - with the matrix polymer to achieve a molecular dispersion of the globular molecules. Only in this case, pore sizes below 10 nm might be achieved after complete decomposition of the labile molecules into volatiles (Scheme 1). The matrix polymer has to be temperature stable with no softening while the degradation of the globular molecules takes place. Otherwise, the pores will collapse. For low k materials, the matrix itself should exhibit already dielectric constants below 3 and the content of the pore forming material should exceed 20 %. Furthermore, there exist many other requirements for polymeric materials as dielectric materials in multilayer microelectronic applications which correspond to the conditions of the processing of chips, e.g. high temperature stability, low leakage current, high breakdown voltage, low metal migration into the polymer[11], good adhesion to metal and barrier layers, etching selectivity, low water absorption, chemical mechanical polishing compatibility[12,13] a.o.

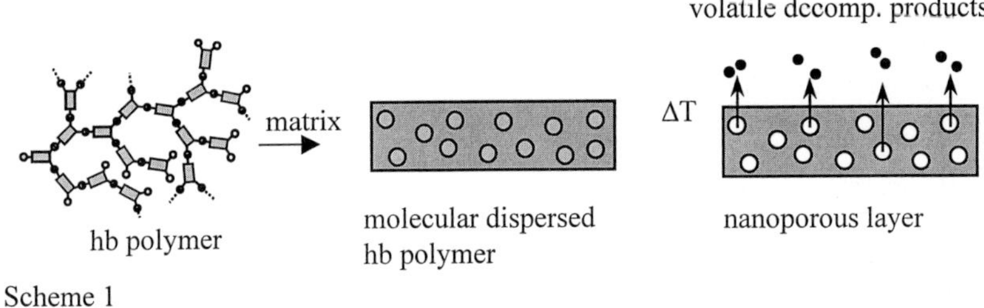

Scheme 1

Results and Discussion

We choose functional hyperbranched polymers containing thermally labile units for this concept. For this, a hyperbranched polymer (HBP) with labile triazene units in each repeating unit (**P1**) was synthesized. Linear polymers containing triazene units in the main chain have proven to be very suitable for polymer ablation processes. They decompose fast upon laser irradiation without leaving solid residues.[14]

For the preparation of the hyperbranched poly(triazene ester)s a low temperature polycondensation of triazene containing AB_2 monomers was performed to avoid undesired early degradation of the labile units. Polymers with molar masses M_n up to 8000 g/mol could be achieved.[15] Scheme 2 shows the structural repeating units.

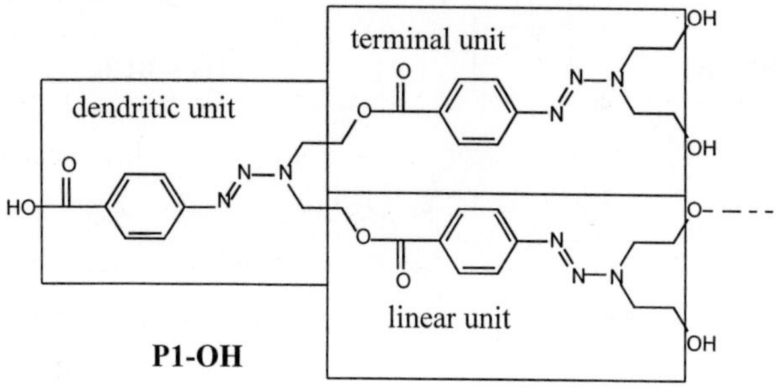

Scheme 2

The structure of the hyperbranched polymers was studied in detail. NMR analysis revealed a statistical degree of branching of 50%. The triazene unit exists in two isomeric forms with the double bond located next to the aromatic ring or next to the aliphatic chain as can be verified by Raman spectroscopy. The polymers can be degraded upon UV irradiation (UV absorption of the triazene functions is in the range of 330 nm) within several minutes or upon thermal treatment. The thermal degradation starts at 160 °C but follows several decomposition steps. About 50% weight loss is observed on isothermal heating at 250 °C after 15 min.

The hyperbranched polymer is very well soluble in polar solvents as expected for a branched structure with polar end groups. However, the high polarity does not allow solubility in solvents also suitable for linear high temperature stable polymers selected as matrix polymers. Therefore, an end group modification was performed to adapt the polarity and solubility and to increase the compatibility of pore forming agent and matrix polymer.

When **P1-OH** is silylated (Scheme 3, **P1-silyl**), one is able to achieve molecular mixtures with different high temperature stable polymer matrixes e.g. a fluorinated polyamideimide PAI or a thermally crosslinkable precursor polymer DVS-BCB (Scheme 3) which is available through Dow Chemicals for application in microelectronics. Figure 1 shows the AFM picture of a homogenous film prepared from PAI and **P1-silyl** (1:1 mixture). The film surface is very smooth and no phase separation can be observed on this scale. After thermal treatment or UV irradiation, surface roughness of films prepared from those mixtures did not increase significantly, however, a small decrease in the refractive index and a change in color of the film could be observed.

150

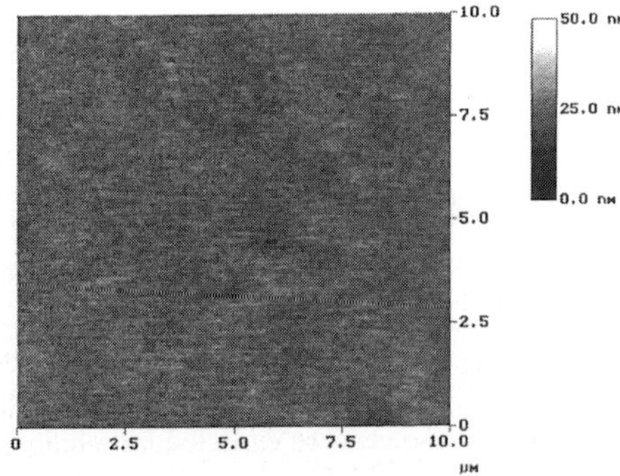

P1-silyl

DVS-BCB

PAI

Scheme 3

Fig. 1: AFM picture (topology, tapping mode) of PAI/**P1-silyl** 1:1, film spin coated from DMF solution (3 wt%).

DVS-BCB crosslinks thermally between 250 and 300 °C (Scheme 4). Films of DVS-BCB with 16% **P1-silyl** have been prepared on silicon wafers by spin coating, dried at 80 °C, annealed at 300 °C for 15 min. under inert atmosphere, and tested for its dielectric properties (Fig.2)[16]. For analysis, the polymer layer (0.5 to 2.5 μm) was prepared on an aluminum (1 μm) coated Si substrate. In Fig. 2 one can observe the upper Al metal layer (1 μm) which has been lithographically structured into squares of different sizes and subsequently etched. The resulting Al squares are necessary for measuring the dielectric constants.

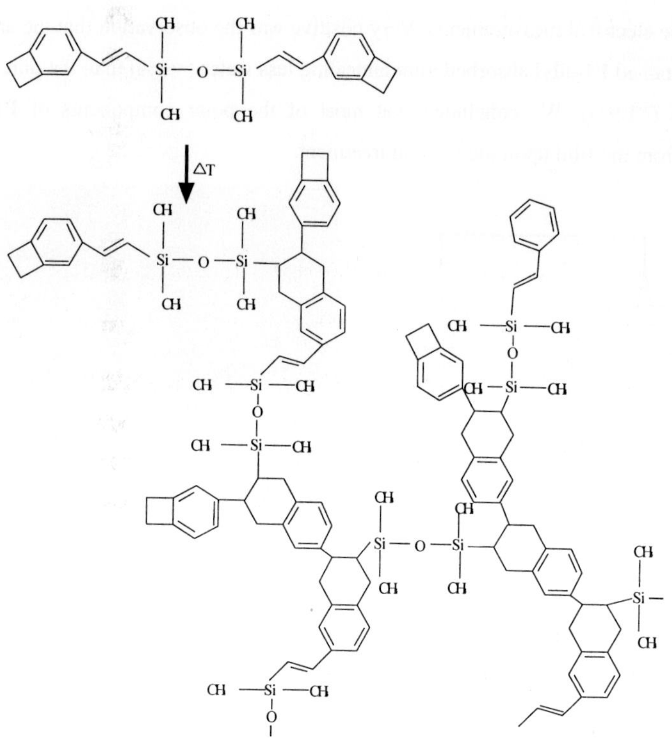

Scheme 4

It is important to note that the crosslinked polymer layer (BCB after removal of the pore forming agent) was compatible with the lithographic process, resisted the metal etching, and showed sufficient homogeneity and adhesion to the metal layers. The second Al-layer was prepared by sputtering after rinsing the crosslinked polymer layer 10 min. with N-methylpyrrolidon. After the lithographic process the Al-etching was performed with a mixture of concentrated phosphoric acid/acetic acid 7:1.

Visually and using light microscopy one could see that the BCB layers formed without pore forming **P1-silyl** had somewhat less defects and a smoother surface. In addition, the films prepared in the presence of **P1-silyl** were thicker (2.6 µm) than pure BCB layers (1.3 µm) due to a difference of viscosity of the spin coating solutions. Both films had a reduction in film thickness of about 11 to 19% after the annealing (crosslinking and pore formation, respectively. The preparation of the resin blend BCB/**P1-silyl** was performed at a standard lab environment. The lack of a special clean room facility explains the significant increase in the defect density of the MIM devices. However the device yield was sufficiently high enough to perform the electrical measurements. Very positive was the observation that the annealed film which contained **P1-silyl** absorbed after annealing less water ($< 1\%$) than the pure crosslinked BCB film ($<1.9\%$). We concluded that most of the polar components of **P1-silyl** were removed from the film upon the thermal treatment.

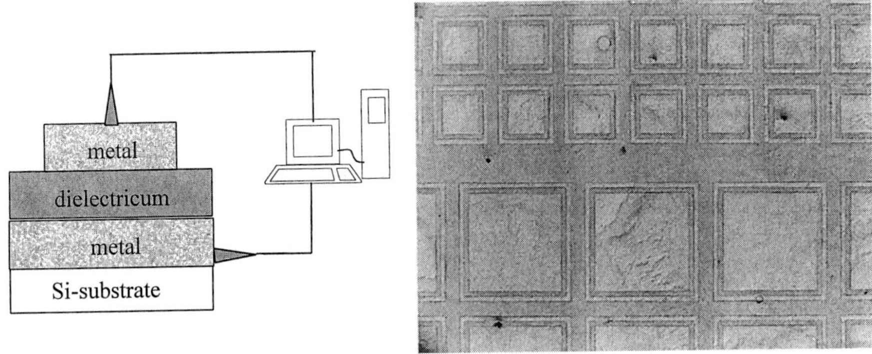

Fig. 2: Schematic representation of an MIM (metal-isolator-metal) structure for measuring dielectric behavior of polymer insulators (left) and the thermally crosslinked (300 $^{\circ}$C, 15 min.) blend of DVS-BCB and 16 % **P1-silyl** in a MIM structure (right, Al squares on top).

In the first experiments measuring dielectric constants a significant decrease of 20 % in ε from 2.7 down to 2.2 of the thermally treated blend with the hyperbranched component compared to pure crosslinked DVS-BCB had been observed without significant changes in surface smoothness or homogeneity of the polymer film on the scale of the light microscope. This points to a successful formation of nanopores in a polymeric matrix. A further hint for some changes in the polymer film after thermal treatment can be found in the AFM pictures (Fig.3). There, a certain inhomogeneity in the cured film can be observed which might point to pores in the bulk phase of the film. In this case, a thinner film on silicon substrate had been prepared via spin coating and this film was cured for in total 90 min. between 140 to 200 $^{\circ}$C. In Fig. 3 the topography as well as the phase sensitive AFM picture is shown. The latter differentiates between softer and harder domains and shows a granular structure with lighter domains indicating harder areas. These harder areas might be caused by increased crosslinking density, and the darker areas can be explained as a softer part of the polymer film covering an air filled pore. The size of the inhomogeneities is in the range of 8-10 nm.

In further studies also a relatively similar values for the breakdown voltage of 5 x 10^{6} and 4 x 10^{6} V/cm was determined for pure DVS-BCB layers and the pore containing material (DVS-BCB + 16wt% **P1-silyl**, cured), respectively. However, the properties of the polymer film are very sensitive to the temperature regime of the crosslinking/decomposition process, thus e.g. the decease of ε could not be verified when the film preparation conditions were changed slightly. A detailed study on verification of the pores and the reduction in dielectric constant as well as the mechanical properties of the new low k material is in progress.

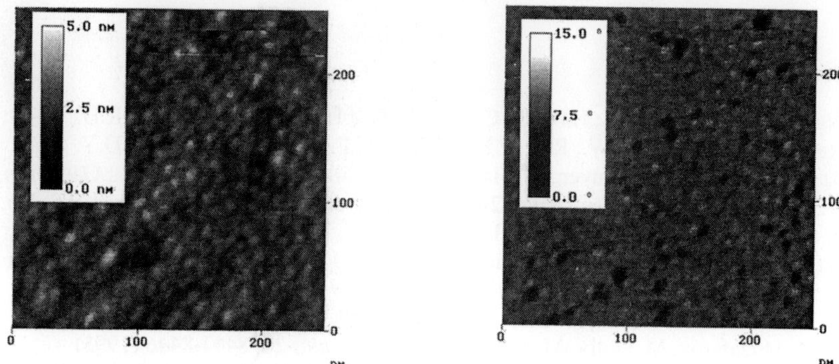

Fig. 3: AFM pictures of the cured film (200 °C, 90 min, 500 mbar) prepared from DVS-BCB + 16 wt% **P1-silyl** (left: topography; right: phase sensitive, picture: 250 x 250 nm).

Up to now, only the crosslinked BCB polymer could be successfully used as matrix for our labile poly(triazene ester) when applied in a microelectronic multilayer system. While also a homogeneous mixture of **P1-silyl** with the polyamideimide (PAI) was obtained, this blend system could not be incorporated into the lithographic process. The adhesion of PAI itself to Al was already lower than that of BCB and after thermal treatment of the blend layer PAI/**P1-silyl**, part of the polymer film fully lost contact to the substrate and measuring ε became impossible.

We can conclude that hyperbranched polymers containing labile groups in the main chain are promising substances as nanopore forming materials in a stable polymeric matrix. Choosing the right matrix and the optimum between crosslinking reaction of the matrix and thermal degradation of the hyperbranched polymer can lead to interesting polymeric dielectrica in microelectronic multilayer systems. In addition, our poly(triazene ester)s allow beside the thermal degradation also a photochemical decomposition. Thus, in case the thermal requirements from the microelectronic processes are lowered in the future, also low temperature pore forming processes are available and therefore polymeric dielectrica with lower thermal stability can be considered.

Acknowledgements

We would like to thank the SMWK (Saxonian Ministry of Science and Art) and the SFB 287 "Reactive Polymers" for financial support of this work. The help of Dr. Komber with the NMR analysis is gratefully acknowledged.

References

1. Nguyen, C.; Carter, K.R.; Hawker, C.J.; Hedrick, J.L.; Jaffe, R.L.; Miller, R.D.; Remenar, J.F.; Rhee, H.W.; Rice, P.M.; Toney, M.F.; Trollsås, M.; Yoon, D.Y. *Chem. Mat.* **1999**, *11*, 3080; Nguyen, C.; Hawker, C.J.; Miller, R.D.; Hedrick, J.L.; Gauderon, R.; Hilborn, J.G.; *Macromolecules* **2000**, *33*, 4281. J.H. Golden, C.J. Hawker, P.S. Ho, *Semiconductor International*, May Issue 2001; http://www.semiconductor.net/semiconductor/issues/issues/2001/200105/02six0105por.asp

2. B. Boury, R.J.P. Robert, R. Nunez, *Chem. Mater.* **10**, 1795-1804 (1998)

3. A.M. Muzafarov, M. Golly, M. Möller, *Macromolecules* **28**, 8444-8446 (1995)

4. Treichel, H.; Withers, B.; Ruhl, G.; Ansmann, P.; Würl, R.; Müller, C.; Dietlmeier, M.; Maier, G. in *Handbook of low and high dielectric constant materials and their applications*, **1999**, p.1-67, Ed. Hari Singh Nalwa, Academic Press, San Diego.; L. Peters, *Semiconductor International,* May Issue 2001, http://www.semiconductor.net/semiconductor/issues/issues/2001/200105/01six0105por.asp

5. J.L. Hedrick, R.D. Miller, C.J. Hawker, K.R. Carter, W. Volksen, D.Y. Yoon, M. Trollsås, *Adv. Mater.* **10**, 1049-1053 (1998)

6. J.L. Hedrick, K.R. Carter, J.W. Labadie, R.D. Miller, W. Volksen, C.J. Hawker, D.Y. Yoon, T.P. Russell, J.E. McGrath, R.M. Briber, *Adv. Polym. Sci.* **141**, 1-43 (1999)

7. K. R. Carter, *Mat. Res. Soc. Symp. Proc.* **476**, 87 (1997)

8. C. Jin, J. D. Luttner, D. M. Smith, T. A. Ramos, *Mat. Res. Soc. Bulletin* **39** (1997)

9. J.L. Hedrick, C.J. Hawker, M. Trollsas, J. Remenar, D.Y. Yoon, R.D. Miller, *Mater. Res. Soc. Symp. Proc.* **519**, 65-75 (1998)

10 J.L. Hedrick, M. Trollsas, C.J. Hawker, B. Atthoff, H. Claesson, H. Heise, R.D. Miller, D. Mecerreyes, R. Jerome, P. Dubois, *Macromolecules* **31**, 8691-8705 (1998)

11. T. Hörnig, K. Melzer, U. Schubert, H. Geisler, J. W. Bartha, Proc. IITC 2000, San Francisco, June 5-7 2000, 211-21312

12. F. Kuchenmeister, U. Schubert, J. Heeg, C. Wenzel, Proc. IITC 1999, San Francisco, CA., May 1999, 158-160

13 F. Kuchenmeister, Z. Stavreva, U. Schubert, K. Richter, C. Wenzel, M. Simmonds, Proc. AMC Tokyo, Japan, oct. 1998

14 O. Nuyken, C. Scherer, A. Baindl; A. R. Brenner, U. Dahn, R. Gärtner, S. Kaiser-Röhrich, R. Kollefrath, P. Matusche, B. Voit, *Prog. Polym. Sci.* **22**, 93 (1997); T. Lippert, J. Stebani, J. Ihlemann, O. Nuyken, A. Wokaun, *J. Phys. Chem.* **97**, 12296-12301 (1993); O. Nuyken, U. Dahn, W. Ehrfeld, V. Hessel, K. Hesch, J. Landsiedel, J. Diebel, *Chem. Mater.* **9**, 485-494 (1997)

15. Eigner, M.; Komber, H.; Voit, B. *Macromol. Chem. Phys.* **202**, 245 (2001)

16 K. Estel, T. Hörnig, C. Wenzel, J. W. Bartha, 1'st International Symp. SFB 287, Dresden, July 16-19 2000

Polymerization of Hydroxymethyloxetanes - Synthesis of Branched Polyethers with Primary Hydroxyl Groups

*M. Bednarek, S. Penczek, P. Kubisa**

Center of Molecular and Macromolecular Studies, PAS,
Sienkiewicza 112, 90-362 Łódź, Poland,
pkubisa@bilbo,cbmm.lodz.pl

Summary: Oxetanes containing hydroxyl groups can conveniently be prepared using commercially available trimethylolpropane or penthaerithritol. The corresponding monomers, such as 3-ethyl-3-hydroxymethyloxetane (EOX) have been polymerized cationically and it has been shown, that resulting polymers are branched. These polymers are of particular interest, since in contrast to some other polyhydroxyethers (like polyglycidol) all of the hydroxyl groups are primary ones. Moreover, all of the $-CH_2OH$ groups are attached to the quaternary carbon atoms. These features are rather unique and may be advantageous from the point of view of possible applications.

Introduction

3-Ethyl-3-hydroxymethyloxetane (EOX) is a monomer combining within one molecule two nucleophilic sites: cyclic ether function and hydroxyl group:

$$O\underset{CH_2}{\overset{CH_2}{\diagdown}}C\underset{C_2H_5}{\overset{CH_2OH}{\diagup}}$$

Several years ago we have studied cationic polymerization of cyclic ethers in the presence of hydroxyl groups containing compounds (alcohols, diols) and we found, that the mechanism of polymerization differs from conventional mechanism of cationic polymerization of cyclic ethers involving successive addition of monomer molecules

CCC 1022-1360/00/$ 17.50+.50/0

156

to the tertiary oxonium ion active centers (Active Chain End Mechanism - ACE).

$$\cdots^{-}CH_2-\overset{\oplus}{O}\!\!<\!\!\rangle_{CH_2} \ + \ O\!\!<\!\!\rangle_{CH_2} \ \longrightarrow \ \cdots^{-}CH_2-O\!\!\frown\!\!CH_2-\overset{\oplus}{O}\!\!<\!\!\rangle_{CH_2}$$

In the presence of sufficiently high concentration of hydroxyl groups containing compounds another mechanism, involving addition of protonated (thus activated) monomer to the terminal hydroxyl groups of the growing macromolecule, is becoming dominating:

$$\cdots^{-}CH_2-OH \ + \ H-\overset{\oplus}{O}\!\!<\!\!\rangle_{CH_2} \ \longrightarrow \ \cdots^{-}CH_2-O\!\!\frown\!\!CH_2-OH \ + \ "H^{\oplus}"$$

The basic features of cationic polymerization of cyclic ethers by Activated Monomer (AM) mechanism have been described in series of papers and reviewed [1-3].

The concept of cationic Activated Monomer polymerization of cyclic ethers has been extended to the system containing both functions (i.e. cyclic ether and hydroxyl group) in the same molecule and it was found, that in the cationic polymerization of glycidol both mechanism compete with each other, leading to branched polymers, having however both primary and secondary hydroxyl groups at the chain termini [4], similar as in anionic glycidol polymerization [5].

More recently we have turned our attention to the group of monomers studied earlier by Vandenberg et al. namely 3-hydroxymethyloxetanes. Vandenberg has already noted that the cationic polymerization of these monomers leads to low molar mass branched polymers [6,7].

In the present contribution, results of our studies of cationic polymerization of 3-ethyl-3-hydroxymethyloxetane (EOX) are discussed. Preliminary results have recently been published [8], almost at the same time results of studies of similar system were published bu Hult et al. [9].

Results and Discussion

Structure of propagating species in the polymerization of EOX

In the previously studied cationic polymerization of glycidol participation of secondary oxonium ions (corresponding to propagation by AM mechanism) and tertiary oxonium ions (corresponding to ACE mechanism) was shown indirectly, by analysing the content of isomerised 1-4-units, which could be formed only by AM propagation step [4]. For four membered rings such an approach is not possible because structure of repeating unit does not depend on the mechanism of propagation step due to the fact that both -CH_2-O- bonds are equivalent. Therefore, the method allowing direct observation of secondary and tertiary oxonium ion active centers, namely phosphine ion trapping method, developed earlier in our group, was applied. This method is based on the transformation of oxonium ions into corresponding phosphonium ions by fast, quantitative and irreversible reaction with tertiary phosphines and determination of resulting tertiary and quaternary phosphonium ions formed respectively from secondary and tertiary oxonium ions by ^{31}P NMR [10].

Using this method it was shown that in the cationic polymerization of EOX oxonium ion active centers are mainly (~90%-mol) secondary oxonium ions although tertiary oxonium ion active species are also present (~10%-mol). The rate constants of reactions involving both types of active centers are not known and therefore their contribution to the chain growth cannot be exactly determined. It may be concluded, however, that suitable conditions for significant participation of secondary oxonium ions in the chain growth do exist in the system.

Reaction of protonated EOX molecule (secondary oxonium ion) with HO- group of growing macromolecule is a source of branching. On the other hand, reaction of tertiary oxonium ion active centers with monomer should in principle lead to linear macromolecules. Thus, high fraction of secondary oxonium ion active species provide a favourable conditions for the formation of highly branched macromolecules.

Molar masses of EOX polymers

Molar masses of poly-EOX were determined by GPC using polystyrene calibration. Because the standard calibration method may be not applicable for determination of correct values of M_n for branched polyethers, the determined values may differ from real M_n values. Thus, M_n's determined by GPC should be rather treated as approximations allowing only relative comparisons.

With this reservation, results of M_n determination for polymers prepared at different polymerization conditions (bulk or solution polymerization at 25^0C), point out to two conclusions:

- M_n is essentially constant throughout polymerization (it does not significantly increase with conversion)

- M_n is considerably lower than it would be expected for polymerization at applied monomer and initiator concentrations if polymerization proceeded without transfer. M_n's irrespectively of polymerization conditions were in the range from 10^3 to $2\cdot10^3$. Polymers isolated at different stages of polymerization (between 20 and 95% conversion) were analysed by MALDI TOF mass spectrometry. The only signals observed in MALDI TOF spectra were the signals of series of macromolecules with molar mass being the multiplicity of molar mass of monomeric units. Results of MALDI TOF analysis, coupled with results of analysis of ^1H NMR spectra indicating the absence of oxetanyl end-groups, led to conclusion that macromolecules do not contain any groups incorporated in initiation and termination. This may occur only if cyclic fragment is formed by intramolecular chain transfer to polymer.

Intramolecular chain transfer to polymer leads to the same structure of cyclic macromolecule irrespectively of mechanism of chain growth as shown in the scheme below. Therefore, structure of isolated polymer does not provide any information on the mechanism of chain growth (AM vs ACE). It explains, however, relatively low molar masses of polymers. and their insensitivity to polymerization conditions. Apparently at the certain stage of the growth of macromolecule (approximately 15 monomeric units), some hydroxyl groups have to appear in so close vicinity of propagating center, that intramolecular chain transfer reaction is becoming inevitable, leading to termination of growth of this particular macromolecule and initiation of growth of another macromolecule.

Below, reactions leading to chain growth and chain transfer, are shown schematically:

Polymerization according to ACE mechanism

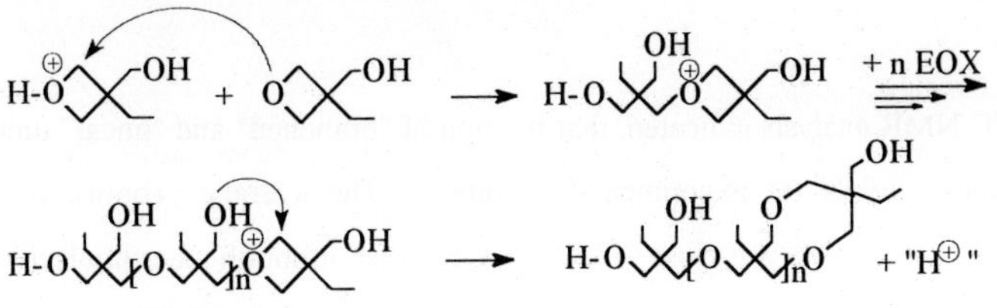

+ n EOX

any HO- group of the macromolecule

Polymerization according to AM mechanism

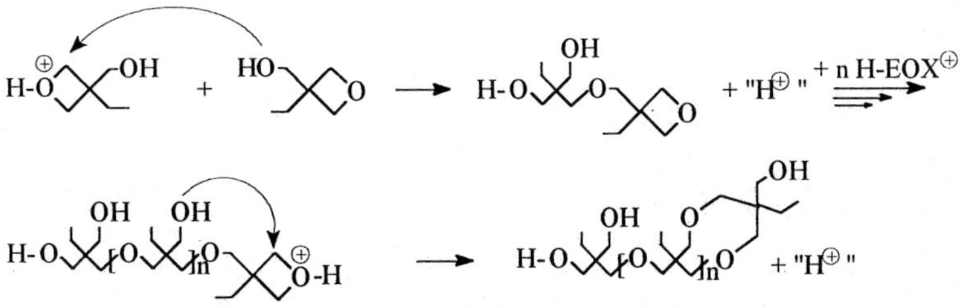

any HO- group of the macromolecule

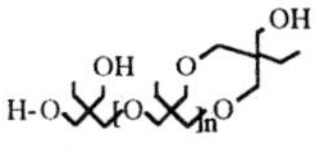

This is only a schematic representation of branched macromolecule containing cyclic fragment formed in reaction between "chain end" and one of the CH_2OH groups.

Microstructure of EOX polymers

Microstructure of EOX polymers may be studied conveniently by ^{13}C NMR. Differences in chemical shifts of quaternary carbon atoms in structural units shown below allowed determination of their fractions in polymers obtained at different experimental conditions:

$$CH_2\text{-}OH \qquad\qquad CH_2\text{-}O\text{-} \qquad\qquad CH_2\text{-}OH$$
$$\text{-}CH_2\text{-}\overset{|}{C}\text{-}CH_2\text{-}O\text{-} \qquad \text{-}CH_2\text{-}\overset{|}{C}\text{-}CH_2\text{-}O\text{-} \qquad \text{-}CH_2\text{-}\overset{|}{C}\text{-}CH_2\text{-}OH$$
$$C_2H_5 \qquad\qquad C_2H_5 \qquad\qquad C_2H_5$$

"linear units"   "branched" units   terminal units

42.67 ppm δ (^{13}C NMR)   41.96 ppm δ (^{13}C NMR)   43.42 ppm δ (^{13}C NMR

Results of ^{13}C NMR analysis indicated, that fraction of "branched" and "linear" units depended only slightly on experimental conditions. The average proportion of "branched" to "linear" units was close to 1:2 indicating, that although macromolecules are indeed branched, macromolecules contain significant proportion of "linear" (not branched) units.

Conclusions

Results presented in this communication may be summarized as follows:

a. In the cationic polymerization of EOX both secondary and tertiary oxonium ion active centers participate in the chain growth, the former ones being much more abundant.

2. Chain growth is terminated after reaching certain stage of growth (in terms of number of monomeric units incorporated into macromolecule) due to intramolecular chain transfer to polymer.

3. Not each of the potential branching points is indeed forming one, on the average only one out of three HO- groups participate in formation of branching point.

4. The fact, that macromolecules, after undergoing intramolecular chain transfer to polymer, do not grow any further by addition of protonated EOX to hydroxyl groups of the macromolecule, although this reaction has to occur at the earlier stage because it leads to formation of branching points, indicate that HO- groups are at certain stage becoming unaccessible for reaction. This is tentatively explained by strong aggregation of HO- groups due to hydrogen bonding, which may lead to particular conformation in which aggregated HO- groups are embedded inside the macromolecular coil.

Basing on these conclusions, in order to prepare polymers of EOX with higher M_n's and higher degree of branching, the following approaches may be envisaged:

1. Increase M_n by copolymerizing EOX with small amount of difunctional monomer giving very similar structure of repeating unit e.g. with dioxetane shown below:

2. Increase the degree of branching by using easily available multihydroxyl initiators/chain transfer agents, such as those shown below:

$$HO\text{-}CH_2\text{-}\underset{\underset{C_2H_5}{|}}{\overset{\overset{CH_2\text{-}OH}{|}}{C}}\text{-}CH_2\text{-}OH \qquad HO\text{-}CH_2\text{-}\underset{\underset{C_2H_5}{|}}{\overset{\overset{CH_2\text{-}OH}{|}}{C}}\text{-}CH_2\text{-}O\text{-}CH_2\text{-}\underset{\underset{C_2H_5}{|}}{\overset{\overset{CH_2\text{-}OH}{|}}{C}}\text{-}CH_2\text{-}OH$$

$$HO\text{-}CH_2\text{-}\underset{\underset{CH_2\text{-}OH}{|}}{\overset{\overset{CH_2\text{-}OH}{|}}{C}}\text{-}CH_2\text{-}O\text{-}CH_2\text{-}\underset{\underset{CH_2\text{-}OH}{|}}{\overset{\overset{CH_2\text{-}OH}{|}}{C}}\text{-}CH_2\text{-}OH$$

3. Increase the degree of branching by reducing the possibility of aggregation of HO-groups i.e. by using solvent or additive known to reduce the extent of intramolecular hydrogen bonding.

All of these approaches are presently under investigation.

[1] K. Brzezińska, R. Szymański, P. Kubisa, S. Penczek, *Makromol. Chem., Rapid Commun.,* **1986**, *7*, 1

[2] S. Penczek, P. Kubisa in *Ring-Opening Polymerization,* Brunelle, D.J., Eds.; Hanser Publishers: Munich, 1993; p 13.

[3] P. Kubisa, S. Penczek, *Prog. Polym. Sci.,***1999**, *24*, 1409

[4] R. Tokar, P. Kubisa, S. Penczek, A. Dworak, *Macromolecules* **1994**, *27*, 320

[5] B. Voit, *J. Polym. Sci., Polym. Chem. Ed.* **2000**, *38*, 2505

[6] E. J. Vandenberg, *J. Polym. Sci., Polym. Chem. Ed.* **1985**, *23*, 915

[7] E. J. Vandenberg, J. C. Mulis, R. S. Juvet, *J. Polym.. Sci., Polym. Chem. Ed.* **1989**, *27*, 3083

[8] M. Bednarek, T. Biedroń, J. Heliński, K. Kałużyński, P. Kubisa, S. Penczek, S. *Macromol. Rapid Commun.* **1999**, *20*, 369

[9] H. Magnusson, E. Malmström, A. Hult, A. *Macromol. Rapid Commun.* **1999**, *20*, 453

[10] K. Brzezińska, W. Chwiałkowska, P. Kubisa, K. Matyjaszewski, S. Penczek, S. *Makromol. Chem.* **1977**, *178*, 2491

Acknowledgment: The support of KBN grant 7 T09A 011 20 is acknowledged

Amphiphilic Poly(oxazoline)s – Synthesis and Application for Micellar Catalysis

Oskar Nuyken, Peter Persigehl and Ralf Weberskirch*

Lehrstuhl für Makromolekulare Stoffe, TU München
Lichtenbergstr. 4, 85747 Garching/ Germany
Tel.: 089/28913570, Fax: 089/28913562

Introduction

Homogeneous and heterogeneous catalytic processes compete on industrial scale for more then three decades [1]. The advantages of homogeneous catalysis are beside others high activity, mild reaction conditions, high selectivity and no diffusion problems. The disadvantages of such systems include difficulties with recycling of the catalyst and product separation [2].

Table 1: Characteristics of homogenous versus heterogenous catalysis.

	homogeneous catalysis	heterogenous catalysis
activity (with respect to metal content)	high	variable, only on surface
reaction conditions	mild	harsh
catalyst life time	variable	long
catalyst recycling	expensive	not necessary
mechanistic understanding	In situ spectroscopy	seldom
selectivity	high	variable
diffusion problems	no	possible

Possibilities to combine the benefits of both homogeneous and heterogeneous catalysis are liquid-liquid two phase systems [3] but also micro heterogenization via colloidal assemblies [4].

                     CCC 1022-1360/00/$ 17.50+.50/0

Aqueous biphasic catalysis represents the most important special case of liquid-liquid biphasic processes and has revolutionized the catalytic process methodology allowing the separation of reaction products from the homogenous catalyst phase by simple phase separation (decantation) after the reaction is complete to recycle the catalyst [5]. Due to the physiological, economical and the safety related process engineering in combination with chemical and physical properties of water, aqueous biphasic catalytic processes gained enormous importance over the last 15 years in many industrial processes. Some examples are summarized in the following Table 2.

Table 2: Industrial processes of two-phase catalysis with water as one phase.

process/products	catalyst	capacity (t/a)	company
Hydrodimerisation butadiene → octadienol	Ru/TPPTS	5000	Kuraray
selective hydrogenation → unsaturated alcohols	Ru/TPPTS	----	Rhone-Poulonc
C-C coupling → geranylacetone	Ru/TPPTS	> 1000	Rhone-Poulonc
hydroformylation → butanal, pentanal	Ru/TPPTS	> 300000	Hoechst AG Union Carbide
SHOP → C_{12}-C_{20} α-olefines	Ni/div.	> 10^6	Shell

Hydroformylation is a major industrial process producing aldehydes from olefins, carbon monoxide and hydrogen in the presence of a rhodium catalyst combined with phosphorous ligands.

Scheme 1

The industrial process of the hydroformylation was developed by the Ruhr-Chemie AG [6] using the highly water-soluble catalyst RhH(CO)(tppts)₃ in a two-phase system. Although short chain olefins could be successfully converted to their corresponding aldehydes [7], the catalyst showed low activity in the hydroformylation of higher olefins, such as 1-octen. The major reason for that is the limited mass transfer as a result of low water solubility of such long alkyl derivatives in water [8].

Several approaches have been used to improve the reaction rate including surfactants [9], stablilization of the catalytic groups at the liquid-liquid interface with promoter ligands [10] and supported aqueous phase catalysts [11]. Problems with product isolation and catalyst recycling, however, limit the scope of these approaches.

Our approach to solve the problem of limited mass transfer of hydrophobic reactants is the use of block copolymer forming micelles. It's well recognized in all branches of chemistry that the role of a chemical reaction can be very sensitive to the nature of the reactant environment. The presence of micellar aggregates can provide therefore a beneficial effect through the potential to solubilize a reactant that would otherwise not have significant solubility in the reaction media. To prepare a catalytically active micelles for the hydroformylation process of long chain olefins, the catalyst has to be part of the hydrophobic core, as well.

Micelles, microscopically dispersed in the homogenous aqueous phase, that are able to solubilize both, the hydrophobic reactant and the catalyst represent therefore a promising system for homogenous two-phase catalysis.

Idea of the catalytically active micelle
- hydrophobic core replaces one solvent phase
- solubilization of unpolar substrates
- fixation of catalyst

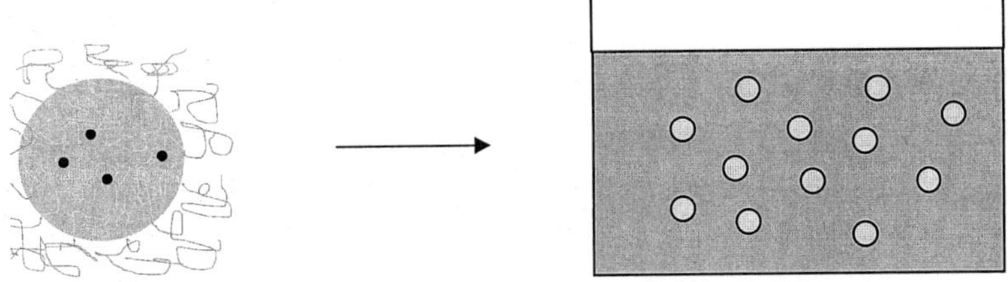

Scheme 2: Micellar catalysis as a approach for homogenous two-phase catalysis

In this presentation, micellar catalysis is applied to the hydroformylation of 1-octene:

Synthesis of amphiphilic block copolymers

Choice of monomer system

The cationic ring-opening polymerisation of 2-oxazolines provides among many others an excellent methodology for the design of amphiphilic block- and graft copolymers [12] and the easy introduction of functional end-groups via controlled initiation or termination [13]. In addition the synthetic versatility of the 2-oxazoline monomer system offers the possibility to control the polarity of the resulting polymer from hydrophilic to lipophilic and fluorophilic.

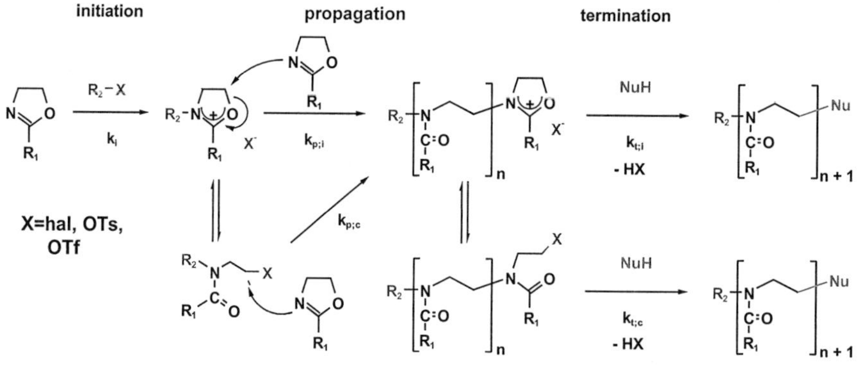

Scheme 3: Mechanism of 2-oxazoline polymerization

The introduction of suitable ligands which are able to immobilize attractive transition metals, e.g. Rh for the hydroformulation is a critical step for the preparation of a micellar catalyst.

An amphiphilic polyoxazoline with triphenylphosphine moieties could become available, i) via sequential monomer addition techniques of suitable monomers, ii) by introducing the desired functional group with the terminating agent or iii) via controlled initiation, using initiators with carry triphenylphosphine units.

We chose the first route, since it offered the best control over metal ligand concentration in the miceller system. The synthesis of the oxazoline monomer with a triphenyl side chain and it's subsequent incorporation in amphiphilic blockcopolymers is depicted in scheme 4.

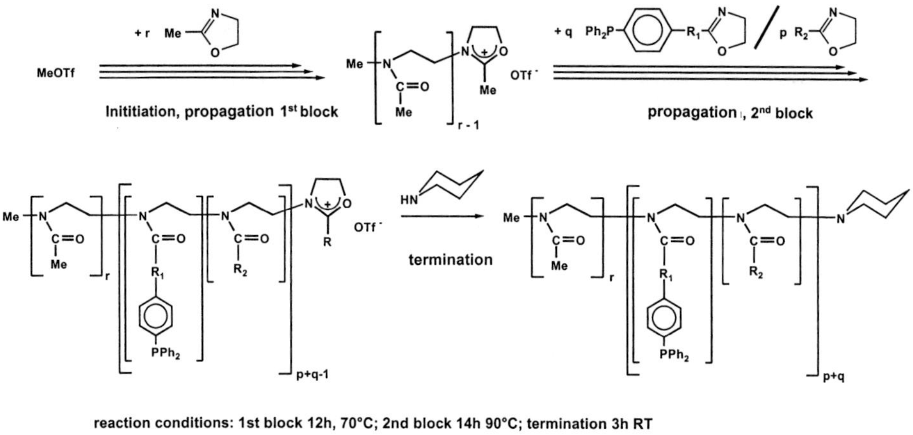

reaction conditions: 1st block 12h, 70°C; 2nd block 14h 90°C; termination 3h RT

Scheme 4: Synthesis of amphiphilic block copolymers by sequential monomer addition

It turned out, however, that the triphenylphosphine units are not inert under conditions which are necessary for the ring-opening polymerization of the oxazoline. Model studies indicated almost complete alkylation oft the triphenylphosphine side chains by the growing cationic chain end, see figure 1.

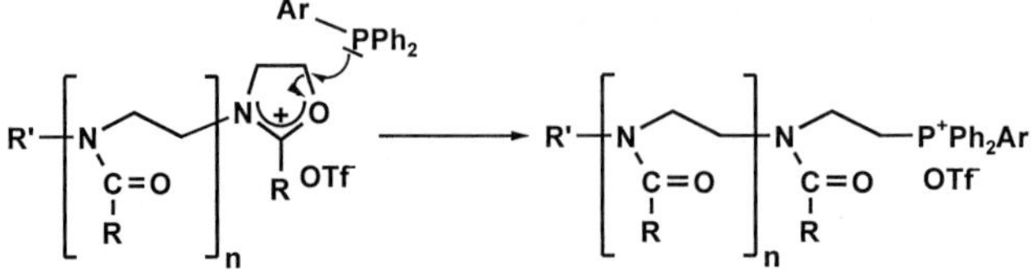

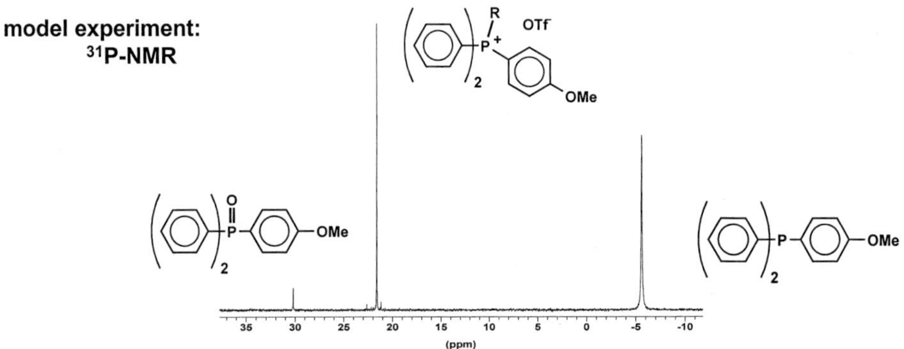

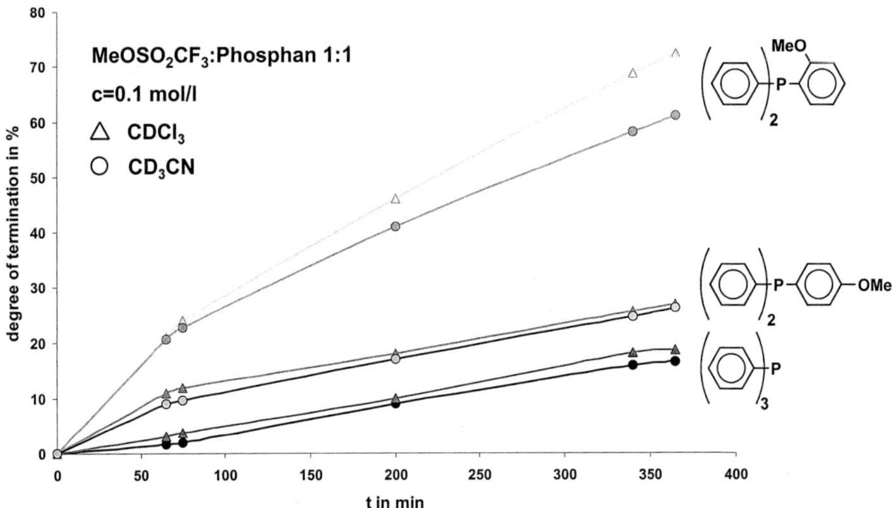

Figure 1: Study of the termination reaction of methyltriflate with different phosphane derivatives

Functionalization of a polymer precursor

A possibility to overcome the problem is the post-analogues incorporation of the triphenylphosphine units using Pd catalysed P-C coupling of diarylphosphine units with aryliodide polymer side chains.

Model studies showed, that the palladacyclus developed by Herrmann et al. [14] showed high activity and selectivity

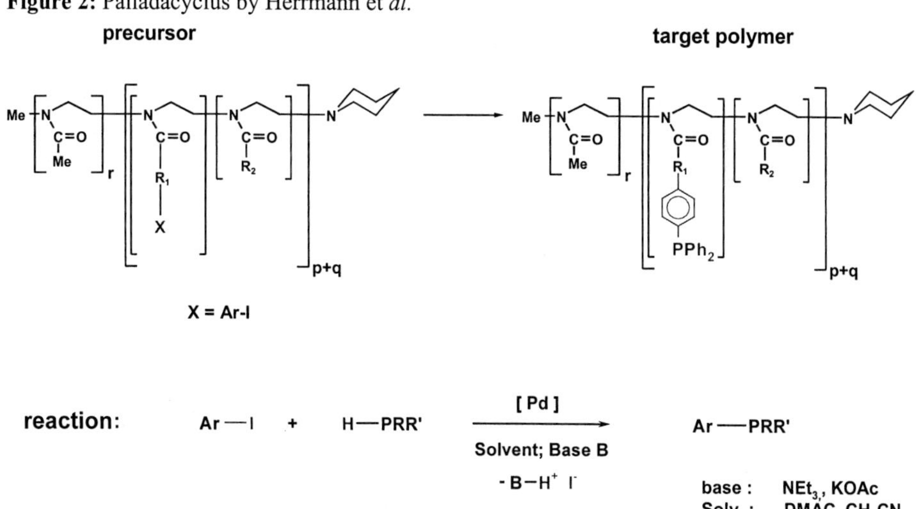

Figure 2: Palladacyclus by Herrmann et *al.*

Scheme 5: Precursor route for the synthesis of functionalised amphiphilic polymers

At least three different routes were developed for the synthesis of aryliodine substituted 2-oxazolines. The pathway depicted in scheme 6 was found to be most suitable and was preferred in the following studies:

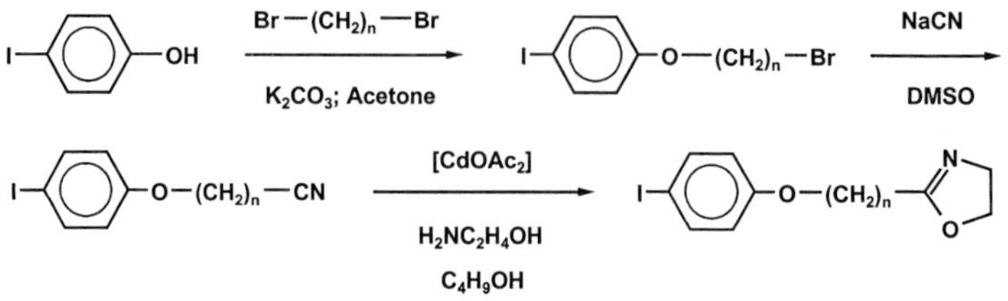

Scheme 6: Synthesis of the aryliodide containing oxazoline monomer

The complete scheme for the synthesis of the precursor polymer is shown in the following scheme 7. First, 2-Methyl-2-oxazoline is initiated with MeOTf – the length of the resulting water-soluble poly(2-methyl-2-oxazoline) was controlled by the ratio of [M] : [I]. After the complete conversion of the first monomer a mixture of the aryliodine derivative and an alkyl substituted oxazoline was added and copolymerized. Finally, the growing chains were terminated by the addition of piperidine.

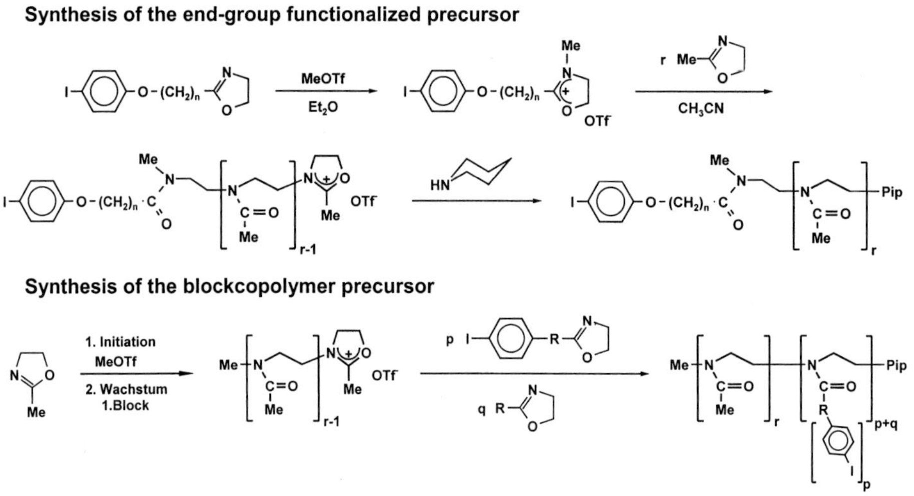

Synthesis of the end-group functionalized precursor

Synthesis of the blockcopolymer precursor

Scheme 7: Complete scheme of the precursor polymer synthesis

The polymer modification gave the desired polymers in almost quantitative yield.

Table 3: Results of the polymer modification by post-analogue Pd-catalysis.

Solv.	[Pd]	ε	τ	t in h	c in %	remarks
CH$_3$CN	Pd(OAc)$_2$	100	2,5	14	85	side products
DMAc	Pd(OAc)$_2$	100	2,5	14	90	side products
CH$_3$CN	1	50	2,5	14	>95	----
CH$_3$CN	1	100	2,5	10	>95	----
CH$_3$CN	1	250	1,2	48	>95	----

ε=n(Ar-I):n(Pd); τ=n(HPPh$_2$):n(Ar-I)

Hydroformylation of 1-octene

As already mentioned, for an industrial application of the two phase catalysis towards hydroformylation it is essential that the educt (olefine) is reasonable good soluble in water. Therefore, usually higher olefins can not be converted into the corresponding aldehydes by a common two phase hydroformylation process due to their poor solubility in water. The strategy of this work is shown in the following scheme using micelle formation of amphiphilic blockcopolymers with triphenylphosphine ligands covalently linked to the hydrophobic part of the polymer backbone.

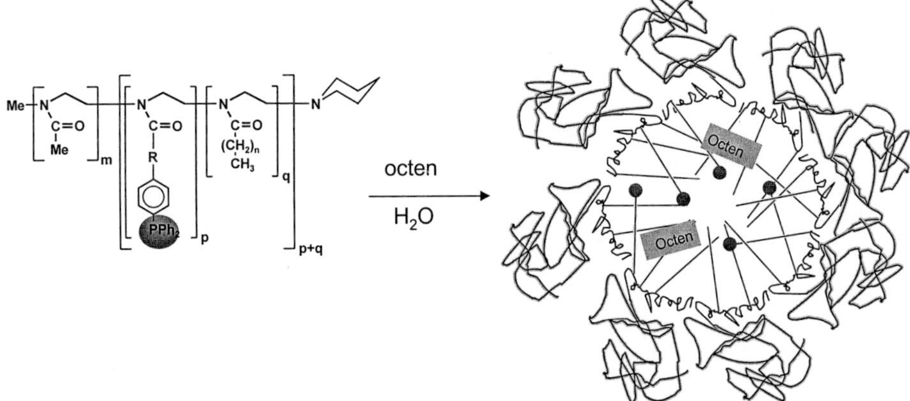

Scheme 7: Micelle formation of amphiphilic blockcopolymers

Above the critical micelle concentration, the amphiphilic blockcopolymers start to form micelles, which have a hydrophobic core where the phosphine ligands and consequently the transition metals are immobilized and a hydrophilic shell to guarantee water-solubility. Since

1-octene is good soluble in the micellar core, its local concentration close to the catalytic center is high, which is essential for its efficient hydroformylation.

The following table summarizes experimental data for the hydroformylation of 1-octene under different experimental conditions. The results in the presence of TPPTS (run 1), SDS/TPPTS (run 2) and SDS/PPh$_3$) (run 3) indicate low catalytic activity with almost no product formation and TON values ranging from 20-100 and TOF values between 2 and 10.

The catalytic results improve dramatically in the presence of the prepared polymeric macro ligands. Product yield increases up to 81 % with TON and TOF values of up to 4650 and 1610, respectively.

Table 4: First results of the hydroformylation in the presence different catalyst systems.

run	catalyst	P:Rh	Oc:Rh	t in h	yield	c in %	n/iso	TON	TOF
1	TPPTS	25	2000	10	<1%	<5	73:27	20	2
2	SDS, TPPTS	25	2100	10	<1%	<6	72:28	64	6
3	SDS, PPh$_3$	25	2000	10	<1%	<8	73:27	100	10
4	Me$_{30}$IPP7$_4$Hep$_2$	29	2000	14	24%	99	80:20	480	34
5	Me$_{30}$IPP7$_2$Hep$_4$	22	7400	4	81%	95	76:24	4650	1610

c(Phos.) = $1.0*10^{-3}$ for run **1, 2, 3, 5** and $2.5*10^{-3}$ for run **4**; standard: IO

Summary

Novel amphiphilic blockcopolymers for micellar catalysis applications were prepared by living cationic polymerisation of 2-oxazoline monomers. Phosphine ligands suitable for transition metal immobilization were introduced by palldium catalyzed post analogue P-C coupling of aryliodide containing polymers with bisaryl phosphine.

First results of the 1-octene hydroformylation indicate high catalytic activity of the micellar catalyst. This supports our idea that this system functions as a nanoreactor and should allow us to extend this strategy also to other catalytic processes.

1. G.M.Parshall, S.D.Iffel, *Homogeneous Catalysis 2.Ed*. J.Wiley, New York 1992

2. B. Cornils, *Top. Curr. Chem*. 206, 133 1999

3. J. Manassen in: F. Barsolo, R. L. Burwell, *Catalysis in Research*, Plenum, London p. 183 (1973)

4. J. H. Fendler, E. J. Fendler, *Catalysis in Micellar and Macromolecular Systems*, Academic Press, London, 1975.

5. B. Cornils, W. A. Herrmann, *Aqueous phase Organometallic Catalysis – Concept and Application*, Wiley-VCH, Weinheim.

6. W. A. Herrmann, C. W. Kohlpainter, Angew. Chem. Int. Ed. 32, 1524 (1993)

7. E. G. Kuntz, Fr. Patent 2 314 910, Rhone-Polounc Recherche (1975)

8. T. Bartik, B. B. Bunn, B. Bartik, B. E. Hanson, Inorg. Chem. 33, 164 (1994)

9. G. Oehme, *Micellar Systems* in Ref. 5

10. V. Chaudari, B. M. Bhanage, R. M. Deshpande, H. Delmos, Nature 373, 501 (1995)

11. W. T. Ford, CHEMTECH 7, 436 (1984)

12. S. Kobayashi, Prog. Polym. Sci. 15, 751 (1990)

13. R. Weberskirch, J. Preuschen, H. W. Spiess, O. Nuyken, Macromol. Chem. Phys. 201, 995 (2000)

14. W. A. Herrmann, C. Brossmer, K. Öfele, C. P. Reisinger, T. H. Priermeier, M. Beller, H. Fischer, Angew. Chem. 107, 1989 (1995)

Macromol. Symp. **177**, 175–183 (2002)

Monolayers and Membranes from Amphiphilic Polymers

*H. Xu[1], R. Heger[2], F. Mallwitz[1], M. Blankenhagel[2], C. Peyratout[2], and Werner A. Goedel[1] **

[1] Macromolecular and Organic Chemistry, OC3, University of Ulm, Germany
[2] Max-Planck-Institut f. Kolloid- & Grenzflächenforschung, Berlin, Germany

Summary: Nanometer thin, elastomeric membranes with considerable application potential in micro mechanics and materials science can be prepared by transferring monomolecular layers of polymers with ionic head groups from the water surface to solid substrates with holes. If monolayers of liquid polymers are transferred to substrates with openings they initially cover the openings, but finally rupture within a couple of minutes after transfer. However, if the polymer monolayers are stabilised by vitrification, chemical or physical cross-linking, they can be transferred to cover openings in solids substrates as stable membranes. Especially if monolayers of low glass transition polymers are cross-linked, elastomeric membranes are obtained, which might find application in micro mechanical devices like membrane valves and pumps. Incorporation of either a second, incompatible polymer or hydrophobised colloids leads to laterally structured and porous membranes.

Introduction

The Langmuir-Blodgett (LB) technique offers the opportunity to generate suspended membranes by assembling a monolayer at the air-water interface and transferring it to cover a hole in a solid substrate. However, the preparation of suspended membranes via LB-transfer is generally more difficult than LB-transfer of thin organic coatings onto continuous smooth surfaces: Because it is not supported by an underlying substrate, the suspended membrane itself must be tough enough to withstand mechanical stress during fabrication and final use. Monolayers that are made from low molecular weight compounds or from liquid polymers easily rupture during transfer across a hole. Suspended membranes have been fabricated using glassy polymers, often stabilised by cross-linking [1] [2]. These membranes are usually rigid. For certain applications, like membranes in micro mechanical valves and pumps, it might be advantageous to have elastomeric thin membranes available and to take advantage of the comparatively large reversible deformation of these materials.

176

Membranes made from monolayers of anchored Polymers

When applied to a water surface, liquid polymers like perfluoropolyethers [3], polyisoprene [4], polybutadiene [5] or polydimethylsiloxane [6] with ionic head groups easily form smooth and continuous monolayers. By variation of the polymer chain length and surface concentration, the thickness of these monolayers can easily be tuned in the range of 10 to 100 nm thickness [7]. If these monolayers are transferred to substrates with openings (e.g. an electron microscopy grid) as schematically depicted in Fig. 1 they initially cover the openings as approximately 50 nm thin bilayers. However, these membranes rupture within minutes after transfer. As an example two snapshots of a membrane made via LB-transfer of polyisobutene with a single head group and a chain length of 300 repeat units is shown in Fig. 2. Within 30 minutes, all membrane-covered holes in the grid rupture.

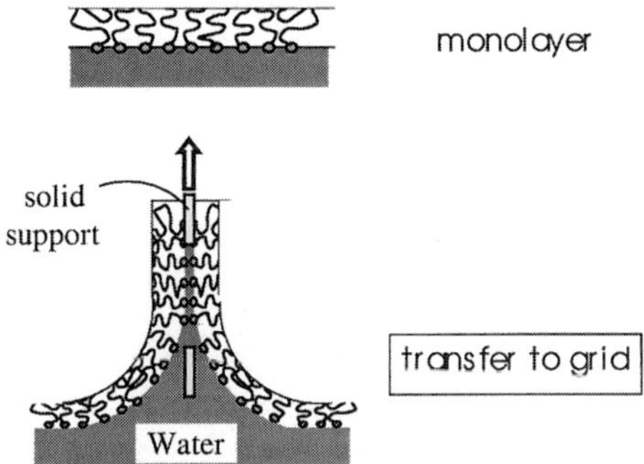

Fig. 1
Scheme of the formation of a freely suspended membrane via Langmuir-Blodgett transfer of an anchored polymer monolayer to substrates with openings.

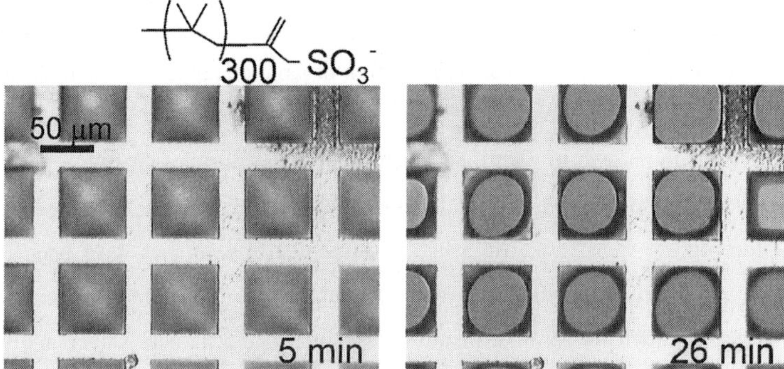

Fig. 2
A freely suspended membrane generated by Langmuir-Blodgett transfer of a monolayer of polyisobutene with a single ionic head group ruptures within 30 minutes. (Light microscopy image with top and bottom illumination.)

This rupture of the membrane can be expected; the membrane closely resembles a soapy membrane made out of a water core coated from both sides with a liquid layer of amphiphiles. Such membranes usually are metastable and rupture, especially if the water of the core region evaporates. However, the rupture can be suppressed and one can obtain stable membranes if the monolayer is solidified before or shortly after transfer. This solidification has been achieved here using three different principles: vitrification, photochemical chemically cross-linking and physical cross-linking.

Freely suspended membranes stabilised by vitrification have been prepared from monolayers of poly-4-*n*-butylstyrene with trimethylammoniumbromide head groups. Polybutylstyrene has a glass transition temperature of 25°C. Hence, at elevated temperatures polybutylstyrene with ionic head groups, applied to a water surface behave essentially like polyisoprenes at room temperature and form smooth and continuous monolayers. Upon cooling the polybutylstyrene monolayers vitrify to room temperature and thus can be transferred to yield solid freely suspended membranes.[8] (Fig. 3)

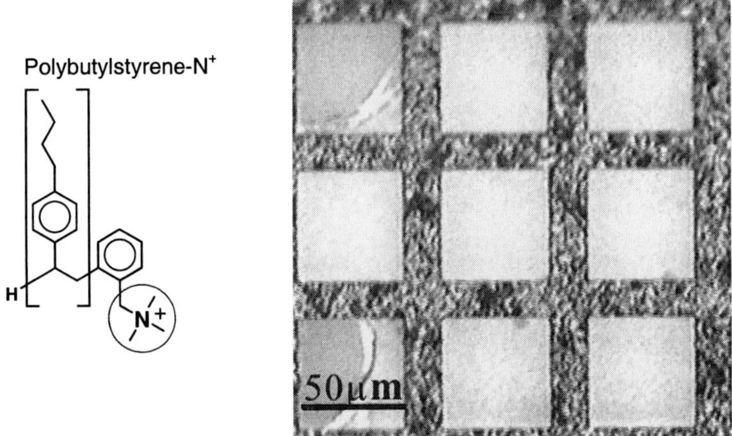

Fig. 3
Light microscopy image of an approximately 20 nm thick freely suspended membrane of polybutylstyrene-N$^+$ prepared by spreading at 40°C, cooling to 10°C and transfer to an electron microscopy grid.

Freely suspended membranes stabilised by chemical cross-linking have been obtained by irradiation of monolayers of polyisoprenes with ionic head groups and anthracene side chains. Upon irradiation with soft UV-light, the anthracene side chains dimerise (see Fig. 4 b). This dimerisation of side chains gives rise to permanent cross-linking points. Since the polyisoprene chains have a low glass transition temperature, this cross-linking transforms the initially liquid monolayer into a thin layer of an elastomer. This layer can easily be transferred across openings in solid substrates.[9] The resulting freely suspended membranes are long term stable (at least several months).

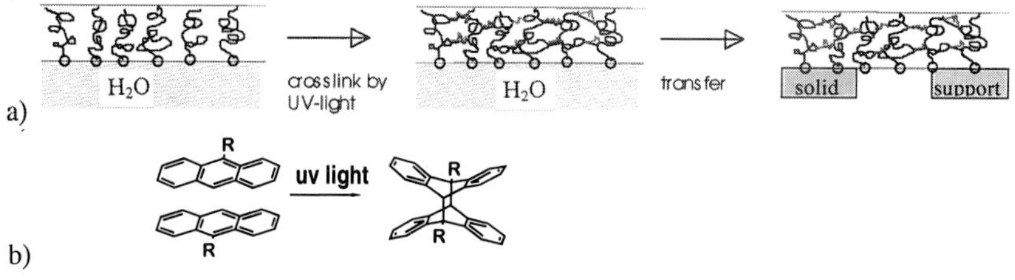

Fig. 4
Scheme of the preparation of elastomeric membranes via cross-linking of the side chains of anthracene tagged polyisopren with ionic head groups.

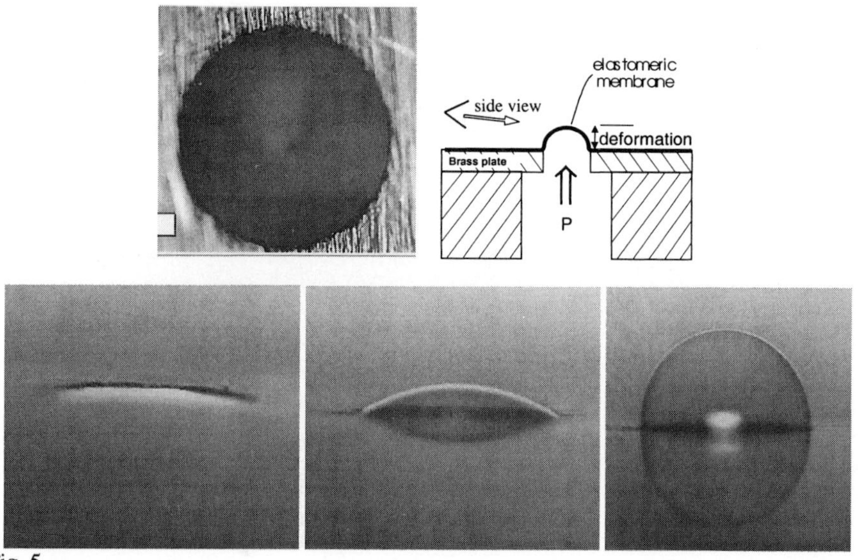

Fig. 5
Top and side view of an elastomeric cross-linked polyisoprene membrane reversibly deformed by a small overpressure from below.

The elastomeric properties of these freely suspended membranes can be shown by applying a small pressure from one side. When the pressure is applied, the membrane bulges. When the pressure is released, the membrane flattens itself reversibly. (see Fig. 5).

In this procedure the monolayer is cross-linked and converted into a solid layer on the water surface; after cross-linking it can sustain neither shear flow nor extensional flow. When a monolayer is transferred to a substrate, which is smaller than the Langmuir trough, it has to undergo two-dimensional flow, otherwise it will develop stress or will wrinkle, especially if several substrates are coated consecutively. Thus, in the experiments depicted above only small substrates were coated and most of the monolayer had to be discarded.

The two dimensional flow problem can be avoided by transferring liquid monolayers and cross-linking them shortly after the transfer. This can be accomplished quite easily by using polymers with more than one ionic group per chain. On the water surface, these polymers behave similar to the polymers with single ionic head groups. Like the linear polymers depicted in Fig. 2, the monolayers of the three arm star polymers can be transferred to cover holes in solid substrates. In the case of the star polymers, however, the ionic groups form inverted micelles when the membrane dries. These inverted micelles efficiently cross-link the polymer and thus give rise to the formation of elastomeric membranes without irradiation being necessary [10] (see Fig. 6).

Fig. 6
Scheme of physical cross-linking of a suspended membrane of polymer chains with multiple head groups. Upon drying of the water core, which is initially present in the transferred membrane, the head groups aggregate and form physical cross linking sites.

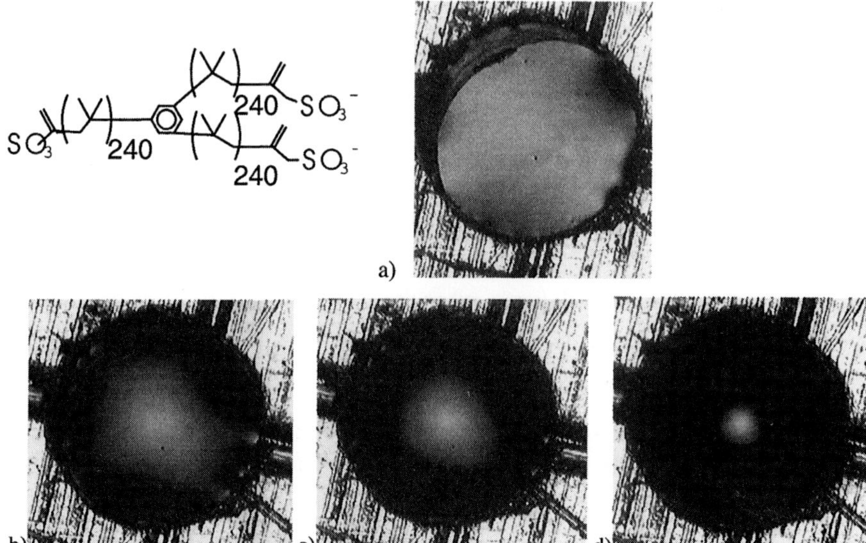

Fig. 7

Top view of an elastomeric polyisobutene membrane cross-linked via aggregation of multiple ionic head groups. a) Membrane spanning a 300 μm hole in a brass plate b) –d) Membrane bulges upward upon applying a small pressure from below.

Like in the case of photochemically cross-linked membranes, these physically cross-linked membranes are elastomeric.   In Fig. 7 b) to d) a continuously increasing pressure of approximately 10-100 Pa is applied to a freely suspended membrane. The higher the applied pressure, the more does the membrane bulge upward. Upon release of the pressure this deformation is completely reversible and can be repeated multiple times.

Mixing the surface active polymers with a second component, e.g. an incompatible second polymer or hydrophobised colloidal particles gives rise to laterally structured monolayers and membranes. For example if anchored polybutylstyrene and anchored polyisobutylene with oppositely charged head groups are mixed and cospread on a water surface, one obtains monolayers with regular lateral pattern with micrometer periodicity.[11] The driving force for this pattern formation is the demixing of the two incompatible polymers, however, the detailed forces determining the morphology and controlling the size of the pattern are currently not completely understood.

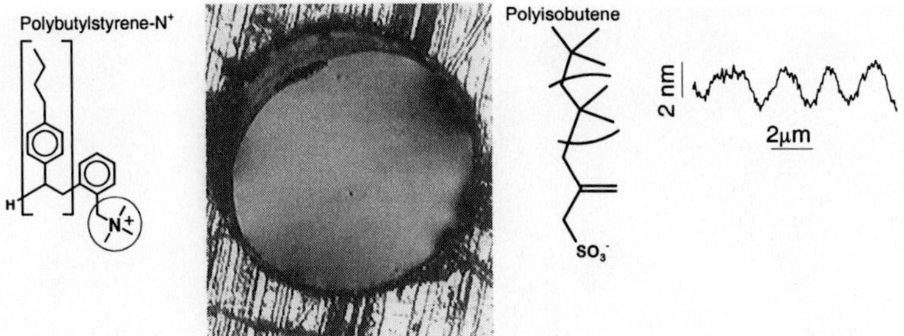

Fig. 8
Scanning force microscopy image and height profile of a mixed monolayer composed of polybutylstyrene with positively charged head groups and polyisobutene with negatively charged head groups.

Cospreading polymers with ionic anchor groups and hydrophobised silica colloids on a water surface, followed by transfer to solid substrates of electron microscopy grids gives rise to mixed monolayers. In these monolayers, domains of silica particles are embedded in a continuous matrix of a polymeric monolayers (see Figure 10 a).

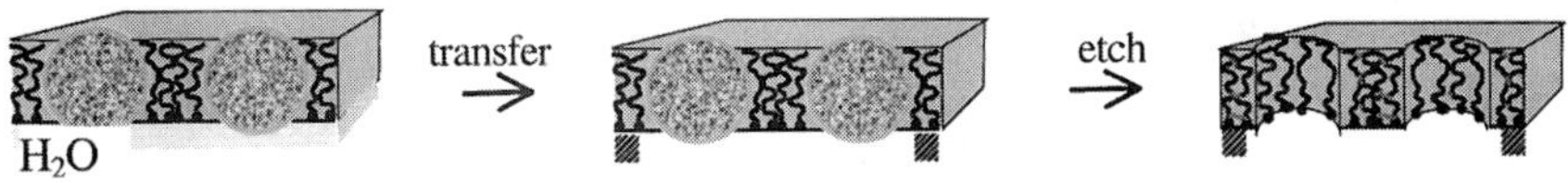

Fig. 9
Scheme of generating porous membranes via incorporation of colloids into polymeric monolayers, followed by cross-linking, transfer and removal of the colloids.

Exposure of these membranes to hydrofluoric acid vapour removes the silica particles. This gives rise to porous monolayers and porous membranes of controlled porosity with a uniform pore size distribution. These membranes are promising for applications like ultrafiltration, bio encapsulation and as masks and moulds for generating new nanoscopic and mesoscopic structures and surface patterns.

Fig. 10
a) hybrid monolayer composed of anchored polymers and silica colloids.
b) porous monolayer obtained after removal of the colloids.

Experimental

Linear polyisoprene with a sulfonate head group and anthracene side groups has been synthesised via living anionic polymerisation followed by platinum catalysed hydrosilylation of the sulfonate terminated parent polyisoprene as published in [4] and [9]; polybutylstyrene with ammonium head group has been synthesised via living anionic polymerisation as described in [8]; polyisobutenes with sulfonate head groups have been synthesised via living cationic polymerisation as described in [12-14]. Monolayers on a water surface were prepared using a 20 cm x 46 cm rectangular Langmuir trough made of polytetrafluoroethylene, equipped with one compression barrier and a floating barrier for the detection of the surface pressure via the Langmuir method (Lauda FW2, Germany). The polymers were usually spread from chloroform solutions which contained 0.05 weight percent of polymer and 10 weight percent of ethanol (polyisoprene, polybutylstyrene) or from 4×10^{-4} weight% solutions in Ethanol/Pentane mixtures (1/50 by weight) (polyisobutenes). UV-illumination was made through the thermostatted, transparent lid of the trough using an array of four 30 cm long fluorescence lamps mounted parallel in an (40×40) cm aluminium housing (Philips TL 36D 25/09N). The emission of the lamps was between 305 to 420 nm, maximum emission was at $\lambda_{max} = 355$ nm. 30 min before and during illumination, the air space above the air-water interface was flushed with nitrogen (5 l/min). Silica colloids coated with polyisobutene amphiphiles (mean radius = 70 nm, polydispersity = 11%, suspended in cyclohexane) were obtained from Utrecht Colloid Synthesis Facility, Van 't Hoff Laboratory for Physical and Colloid Chemistry, Utrecht University, The Netherlands.

Acknowledgements

This work was partially conducted in the Max-Planck-Institute of Colloids and interfaces and in the University of Ulm. The support by H. Möhwald, M. Antonietti, M. Möller, by the Max-Planck Gesellschaft and the Deutsche Forschungsgemeinschaft (Go 693/1, Go 693/6, SFB 569) is gratefully acknowledged. We thank S. Förster (MPI-KGF) for introducing us into living anionic polymerisation and M. Grasmüller and O. Nuyken (TU-Munich) for introducing us to living cationic polymerisation. We thank Dr. Carlos van Kats, and Dr. Judith Wijnhoven, A. van Blaadern, A. Phillipse (Utrecht Colloid Synthesis Facility,) for providing the silica colloids and for helpful discussions.

[1] Seufert, M. Fakirov, C. Wegner, G. *Advanced Materials* **1995** *7* 52-55

[2] Kunitake M., Nishi T., Yamamoto H., Nasu K., Manabr O., Nakashima N. *Langmuir* **1994,** *10*, 3207

[3] W. A. Goedel, C. Xu, C. W. Frank, *Langmuir* **1993**, *9*, 1184

[4] R. Heger and W. A. Goedel, *Macromolecules* **1996**, *29*, 8912

[5] P. Christie, M. C. Petty, G. G. Roberts, Thin Solid Films, 1985, 134, 75

[6] T. J. Lenk, D. H. T. Lee, J. T. Koberstein, *Langmuir* **1994**, *10*, 1857

[7] Baltes H., Schwendler M., Helm C. A., Heger R., Goedel W. A. *Macromolecules* **1997**, *30*, 6633

[8] W. A. Goedel, C. Peyratout, L. Ouali, V. Schädler: *Advanced Materials,* **1999**, *11*, 213-217

[9] W. A. Goedel, R. Heger, *Langmuir,* **1998**, *14*, p. 3470-3474

[10] F.Mallwitz, W.A. Goedel, *Angew. Chemie Int. Ed.* **2001**, *40*, 2557-2557 - *Angew. Chemie* **2001**, *113*, 2716-2718

[11] F.Mallwitz, PhD Thesis, University of Potsdam **1999**

[12] R. Santos, J.P. Kennedy, M. Walters, *Polymer Bulletin,* **1984**, *11*, 261

[13] a) J.P. Kennedy, L.R. Ross, J.E. Lackey, O. Nuyken, *Polymer Bulletin,* **1981**, *4*, 67; b) J.P. Kennedy, L.R. Ross, O. Nuyken, , *Polymer Bulletin,* **1981**, *5*, 5

[14] R. F. Storey, Y. Lee, *J. Polym. Sci. Polym. Chem.* **1991**, *29*, 317

Macromol. Symp. 177, 185–191 (2002)

Shape-Persistent Macrocycles: Building Blocks for Complex Organic and Polymeric Architectures

Sigurd Höger, Klaus Bonrad, Silvia Rosselli, Anne-Désirée Ramminger,*

Thomas Wagner, Beate Silier, Simone Wiegand, Wolfgang Häußler, Günter Lieser,

Volker Scheumann

Max Planck Institute for Polymer Research, Ackermannweg 10, 55128 Mainz, Germany

Summary: Alkyl- and Oligostyrene substituents were attached to shape-persistent macrocycles based on a phenyl-ethynyl backbone. In good solvents for both the rigid core and the flexible corona no aggregation occurred. Whereas, addition of a solvent that selectively solubilizes the corona induced a solvophobic aggregation. For alkyl substituted rings the experimental data were described by a monomer-dimer equilibrium. In contrast, the oligostyrene substituted rings formed more expanded aggregates which were investigated by scattering and by imaging methods. The superstructures are consequently described as hollow supramolecular cylindrical brushes.

Introduction

Shape-persistent macrocycles based on the rigid phenyl-ethynyl backbone are readily obtainable in good to high yields by the oxidative Glaser coupling of the corresponding bisacetylenes. These couplings can be performed either in a statistical approach (Fig. 1a) or by using templates (Fig. 1b)[1].

Attachment of alkyl or oligoalkyl groups to the macrocycles leads to structures that exhibit an essentially two dimensional conformation (Fig. 2). Proton NMR spectra of **1** in THF or CH_2Cl_2 do not show any concentration dependence, and therefore, give no indication for ring aggregation. This result is not surprising because there are no highly polar groups in the molecule that could induce such aggregation. In addition, the size of the π-system relative to the extension of the molecule is rather small, and aggregation of large macrocycles has been reported only for compounds having electron withdrawing groups[2].

CCC 1022-1360/00/$ 17.50+.50/0

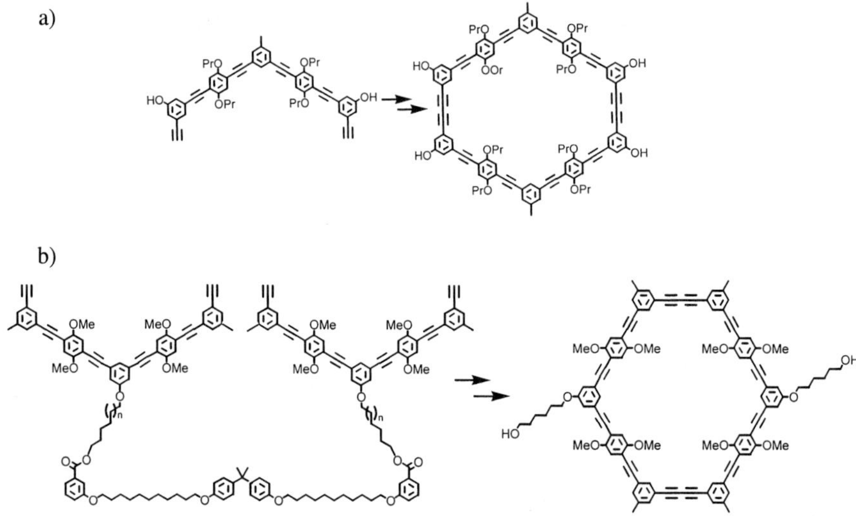

Fig. 1: Synthesis of functionalized shape-persistent macrocycles by the Glaser coupling: a) statistical approach; b) template approach.

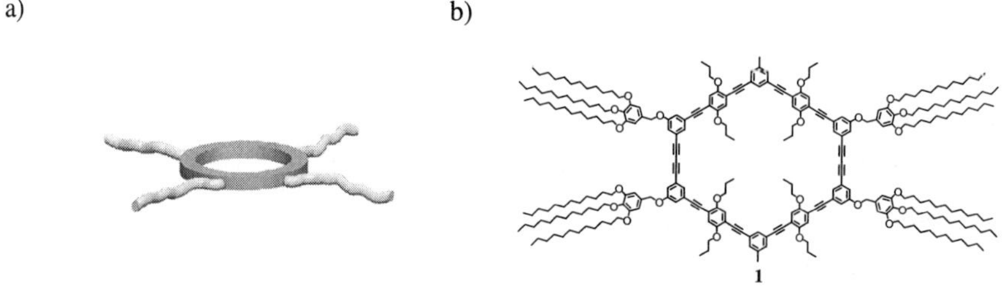

Fig. 2: Oligoalkyl substituted macrocycles: a) schematic; b) structure of **1**.

Macrocycle Aggregation

In a solvent mixture of CH_2Cl_2 and n-hexane, however, a strong concentration dependence of the NMR signals is observed. Assuming a monomer-dimer equilibrium as the predominant association process in solution, the dimerization constant increases from $K_D = 130 \pm 30$ M^{-1} (in CD_2Cl_2/hexane (1:3) at 25 °C) to $K_D = 790 \pm 180$ M^{-1} (in CD_2Cl_2/hexane (1:6) at 25 °C) by increasing the volume fraction of hexane[3]. These data indicate that aggregation of the oligoalkyl substituted macrocycles can be facilitated by the addition of a solvent that acts as a good solvent for the peripheric alkyl chains and as a poor solvent for the core (i.e., solvophobically induced aggregation, Fig. 3)[4]. It must be pointed out, however, that a further increase of the association constants of **1** by increasing the volume fraction of the non polar solvent is not possible due to the limited compound solubility in hexane.

a)   b)

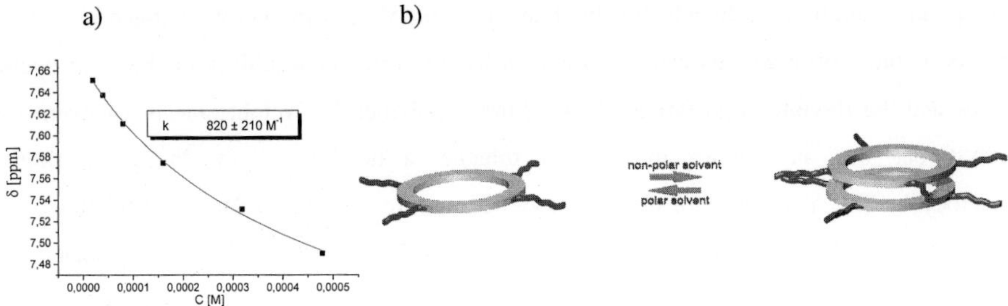

Fig. 3: Aggregation of oligoalkyl substituted shape-persistent macrocycles: a) concentration dependent chemical shift of one aromatic proton of **1**; b) schematic description of a monomer-dimer equilibrium.

A simple way to overcome this problem is the attachment of short, non-crystallizable oligomeric side groups to the ring. DCC-coupling of the corresponding macrocyclic diol and narrow-distributed (PD (M_w/M_n) < 1.1) polystyrene carboxylic acid oligomers of different molecular weight (M_w: 1000(**a**), 1500(**b**), 2500(**c**); 3500(**d**); 5000(**e**) g/mol) produces the products **2a-e** in good yields (70 - 90 %)[5]. The resulting structures can be described as coil-(rigid)ring-coil block copolymers, a subclass of rod-coil block copolymers, which are known to microphase separate to well-ordered superstructures even at relatively small block sizes[6]. Due to the coiled conformation of the side groups, these molecules have an extension in the third dimension (Fig. 4).

a)   b)

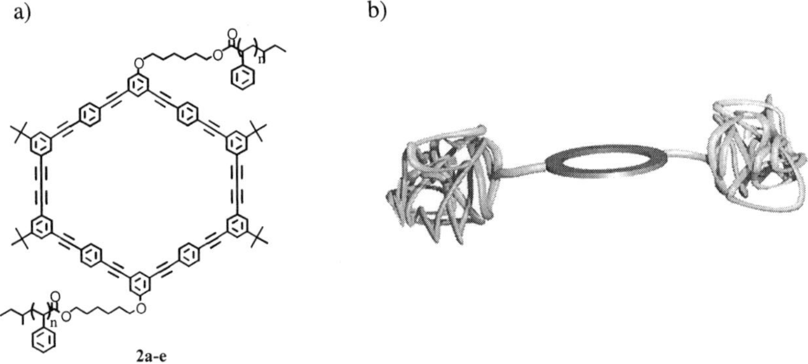

Fig. 4: Oligostyrene substituted macrocycles: a) structure of **2**; b) schematic.

2a-e are readily soluble at room temperature in chloroform, dichloromethane, THF and toluene. In addition **2b-e** are soluble in warm cyclohexane. Upon cooling, **2b** forms a gel at concentrations above 0.5 wt %. Under the same conditions **2c** rapidly forms a very viscous solution, as does **2d** after several days. These solutions are also strongly birefringent with the exception of solutions of **2e**, which exhibit neither unusual viscosity nor birefringence.

As for **1**, these observations can be explained by the different solubility of the rigid cyclic block and the flexible oligostyrene block of the copolymer **2**. Cyclohexane is a solvent for polystyrene but, in contrast to THF or toluene, a non-solvent for the unsubstituted macrocycle. The attachment of the polymeric side groups, even of relatively low molecular weight, is sufficient enough to solubilize the block copolymer in the non-polar cyclohexane at elevated temperatures. Upon cooling, the rigid parts of the block copolymer aggregate if the size of the coiled blocks is below a certain level. As expected, extension and aggregation kinetics strongly depend on the block size of the polystyrene.

Dynamic light scattering (DLS) was performed on solutions of **2c** in toluene and cyclohexane to investigate its aggregation behaviour in more detail. The distribution of the relaxation times in the autocorrelation functions were obtained by CONTIN analysis. As expected, in toluene (0.11 wt %) only one maximum that corresponds to a hydrodynamic radius of approximately 2 nm is found. This maximum corresponds to the size of a simple block copolymer molecule **2c**. In contrary, the light scattering data in cyclohexane at the same concentration showed the formation of more complex structures. The elaboration of the measured autocorrelation function resulted in a distribution of relaxation times that contain two maxima, one at approximately 2 nm and the other at approximately 60 nm (Fig. 5a).

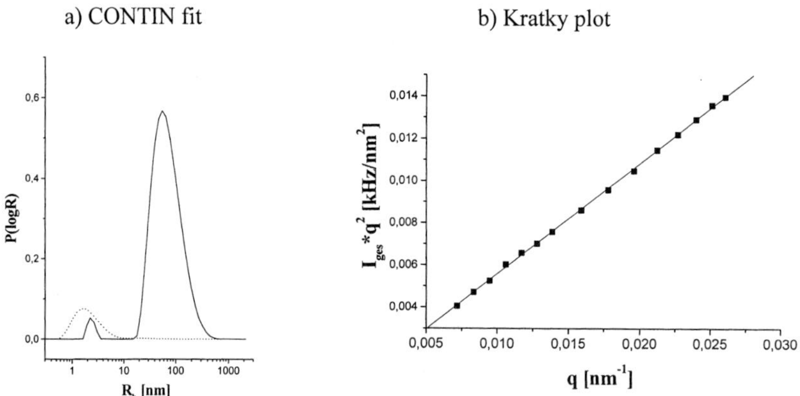

Fig 5: a) CONTIN-fit: rate distribution of **1c** in toluene (⋯⋯) and in cyclohexane (——) at a concentration of 0.11 wt %; b) Kratky plot for the slow mode of **2c** in cyclohexane (■) at a concentration of 0.09 wt % and linear data fit (——).

With increasing concentration the fraction of the larger species increases. The angle dependence of the scattered light intensity of the larger objects (Fig. 2b) suggested the presence of rod-shaped objects with a high virtual persistent length of more than 100 nm and a total length of about 250 nm to 1200 nm[7]. The diameter of these cylindrical objects was determined by ultra small angle X-ray scattering (USAXS) and small angle X-ray scattering (SAXS) in cyclohexane (2 wt %) (Fig. 6).

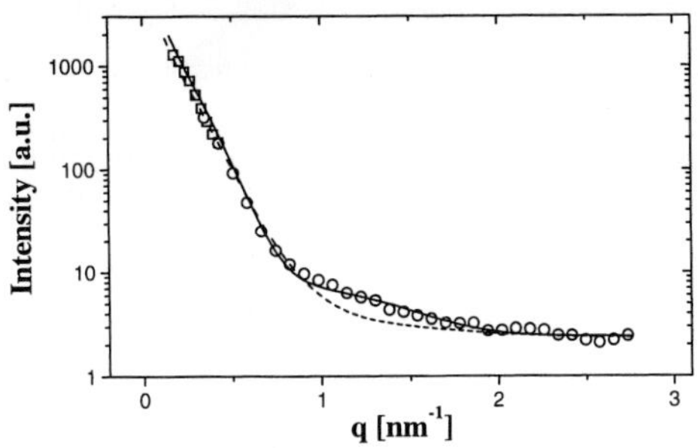

Fig. 6: USAXS (□)-SAXS (O) data of a 2.0 wt % cyclohexane solution of **1c**. The lines represent the form factor for a massive cylinder (----) with a external diameter of 10 ± 3.8 nm, and for a hollow cylinder (——) with an external diameter of 10 ± 3.8 nm and an internal diameter of 1.8 ± 0.5 nm.

The description of the measured data was not possible by using a simple massive cylinder model (showing an abrupt radial electron density change) of any length or thickness. The different length of the oligomeric polystyrene chains and the radial decrease of the cylinder density was therefore taken into account by giving the cylinder a polydispersity with respect to the diameter. The calculated form factor of a massive cylinder 10 nm in diameter having a polydispersity at the outside (d_{out} = 10 nm, polydispersity: 3.8 nm, half width half maximum of a Gaussian distribution) agrees with the observed data at low q-values. However, the observed scattering intensity is remarkably higher than the theoretical prediction in the q-range between 1 and 2 nm^{-1}. Only the use of a radial density profile with a reduced electron density inside the cylinder resolved this discrepancy. The calculated form factor of a cylinder with the same electron density profile at the outside as described before and with an additional reduced electron density at the inside (hollow size d_{in} = 1.8 nm, polydispersity: 0.5 nm, half width half maximum of a Gaussian distribution) fits with the observed data over the entire q-range. The scattering data undoubtedly show that the coil-ring-coil block copolymers aggregate in solution into *hollow* cylinder-shaped objects with a high virtual persistent length (Fig. 7).

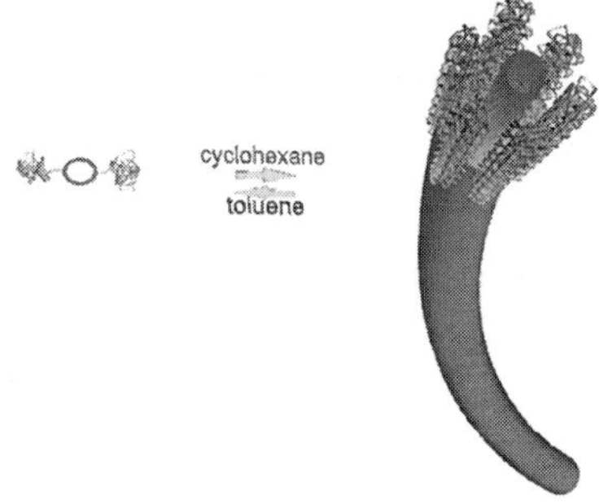

Fig. 7: Schematic aggregation of oligostyrene substituted shape-persistent macrocycles.

Independent evidence of these supramolecular cylinders was obtained by using imaging methods. The transmission electron micrograph (TEM) of a sample obtained by freeze-drying a cyclohexane solution (Pt/C shadowed film) shows ribbons of different width at the sample surface, the narrowest in the range of approximately 15 nm (Fig. 8a).

a) TEM picture

b) AFM picture

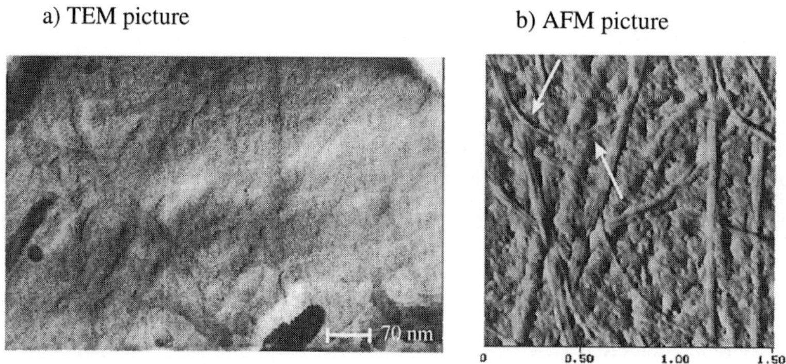

Fig. 8: a) TEM of a C/Pt shadowed film obtained by freeze drying a 0.15 wt % cyclohexane solution of **2c**; b) AFM (amplitude picture, $1.5 \times 1.5 \ \mu m^2$): Film obtained by dipping mica into a 0.15 wt % cyclohexane solution of **2c**.

Atomic force microscopy (AFM) images of a polymer film on mica show long bundles of two or three cylindrical aggregates, together with individual aggregates (Fig. 8b). Taking in consideration the width of the AFM tip, the cylindrical units have a diameter of approximately 10-15 nm. The dimensions obtained by TEM and AFM correspond well with the molecular dimensions obtained by scattering methods and are in accordance with the molecular dimensions of the molecular building blocks[8]. Surveying the most curved cylindrical object found in the AFM images (white arrows, Fig. 8b) and treating it as worm-like chain results in a persistent length of approximately 350 nm of the cylindrical aggregate. The correspondence of these structures with covalently linked polymer brushes suggests that they can be described as hollow supramolecular cylindrical brushes[9].

Conclusion

It has been shown that shape-persistent macrocycles containing solubilizing side groups can aggregate in solvents or solvent mixtures that selectively dissolve the corona of the rings and act as non-solvents for the rigid core. While oligoalkyl substituted rings form predominantly dimers, rings with oligostyrene substituents can aggregate to large hollow supramolecular cylinders. These have been investigated by scattering methods and could be visualized by electron microscopy and atomic force microscopy.

1. Reviews about shape-persistent macrocycles: J. S. Moore, *Acc. Chem. Res.* **30**, 402 (1997); S. Höger, *J. Polym. Sci. Part A*, **37**, 2685 (1999); M. M. Haley, J. J. Pak, S. C. Brand, *Topic Curr. Chem.* **201**, 81 (1999)
2. J. Zhang, J. S. Moore, *J. Am. Chem. Soc.* **114**, 9701 (1992); A. S. Shetty, J. Zhang, J. S. Moore, *J. Am. Chem. Soc.* **118**, 1019 (1996); Y. Tobe, N. Utsumi, K. Kawabata, K. Naemura, *Tetrahedron Lett* **37**, 9325 (1996); Y. Tobe, N. Utsumi, A. Nagano, K. Naemura, *Angew. Chem. Int. Ed. Engl.* **37**, 1285 (1998)
3. R. B. Martin, *Chem. Rev.* **96**, 3043 (1996). The supramolecular polymerization in solution can be described by the same model. Therefore, by analyzing the NMR data alone, we can not discriminate between dimers and polymers of the oligo-alkyl substituted rings
4. S. Höger, K. Bonrad, A. Mourran, U. Beginn, M. Möller, *J. Am. Chem. Soc.* **123**, 5651 (2000)
5. S. Rosselli, A.-D. Ramminger, T. Wagner, B. Silier, S. Wiegand, W. Häußler, G. Lieser, V. Scheumann, S. Höger, *Angew. Chem. Int. Ed. Engl.* in print
6. E. R. Zubarev, M. U. Pralle, L. Li, S. I. Stupp, *Science* **283**, 523 (1999); J. T. Chen, E. L. Thomas, C. K. Ober, G.-P. Mao, *Science* **273**, 343 (1996); S. A. Jenekhe, X. L. Chen, *Science* **283**, 372 (1999); M. Lee, B.-K. Cho, H. Kim, W.-C. Zin, *Angew. Chem. Int. Ed. Engl.* **37**, 638 (1998); M. Muthukumar, C. K. Ober, E. L. Thomas, *Science* **277**, 1225 (1997); H. Engelkamp, S. Middelbeek, R. J. M. Nolte, *Science* **284**, 785 (1999); H.-A. Klok, J. F. Langenwalter, S. Lecommandoux, *Macromolecules* **33**, 7819 (2000)
7. For the estimation of the rod length L from R_h, the relation $R_g/R_h = 2$ was used (valid for rod-like object such as fibrinogen or TMV). With $R_g^2 = L^2/12$ follows $L = [12\ (2R_h)^2]^{1/2}$; H.-G. Elias, *Makromoleküle Bd. 1* (Verlag Hüthig & Wepf, Basel, 1990)
8. S. Höger, V. Enkelmann, *Angew. Chem. Int. Ed. Engl.* **34**, 2713 (1995); S. Höger, V. Enkelmann, K. Bonrad, C. Tschierske, *Angew. Chem. Int. Ed. Engl.* **39**, 2268 (2000)
9. P. Dziezok, S. S. Sheiko, K. Fischer, M. Schmidt, M. Möller, *Angew. Chem. Int. Ed. Engl.* **36**, 2812 (1997); A. D. Schlüter, J. P. Rabe, *Angew. Chem. Int. Ed. Engl.* **39**, 864 (2000)

Elias, Hans-Georg

Makromoleküle

Band 3: Industrielle Polymere und Synthesen, 6. Auflage
Reihe: ELIAS, Makromoleküle (Band 3)

2001. Ca. 600 Seiten,
ca. 400 Abbildungen. Gebunden.
Ca. € 209.-/sFr 354.-/DM 408.77***
Serienpreis: Ca. € 179.-/*
*sFr 309.-/ DM 350.09***
ISBN 3-527-29961-0

Aus dem Inhalt:

STRUKTUR UND EIGENSCHAFTEN VON POLYMEREN/
ROHSTOFFE FÜR POLYMERE/ TECHNISCHE SYNTHESEN/
KOHLENSTOFF-KETTEN/ KOHLENSTOFF-SAUERSTOFF-
KETTEN/ POLYSACCHARIDE/ KOHLENSTOFF-SCHWEFEL-
KETTEN/ KOHLENSTOFF-STICKSTOFF-KETTEN/ PEPTIDE UND
PROTEINE/ ANORGANISCHE POLYMERE

*   €-Preise gültig ausschließlich in Deutschland und Österreich!
** Vorläufige Preise. Änderungen vorbehalten.

Wiley-VCH · 69469 Weinheim · Germany
P.O. Box 10 1161 · Fax +49 (0) 62 01 60 6184
e-mail: service@wiley-vch.de · http://www.wiley-vch.de

5790107L_vi

WILEY-VCH